CONTENTS

THE INTELLIGENCE OF ANIMALS

AND OTHER PAPERS

A THEORY OF LEARNING

by

D.A.O. Williams, MA

David Williams
London
1992

First published 1992 by
David Williams
51 Grange Park Avenue
London N21 2LN

ISBN 0 9519872 0 8

Typeset by Ella Whitehead, Munslow, Shropshire
Printed by The Signal Press Ltd., 137 Great Suffolk Street, London SE1 1PP

Foreword

Natural intelligence may be thought of as a kind of unwritten and unspoken lore or set of beliefs about the world handed down through millions of generations of animals. It is intimately bound up with their capacity to survive. It is quite distinct from experience, which is another important source of information about the environment. An animal's predictions about the consequences of possible future actions depend on both factors.

So far as human beings are concerned the notion that much of our practical everyday knowledge is biologically inherited and not empirical may be hard to accept. I have tried to marshal the arguments in favour of such an unusual epistemological point of view as strongly as I can. Humans are the most intelligent of animals and by equating this hidden store of ideas with the trait I believe I have reached the core of the problem.

Turning now to experience I think most creatures have feelings, many of them not so very different from our own. Sensations and emotions are subject to the same logic which governs the mechanisms behind the learning process. They play a major role in shaping behaviour. The examples I give in the preface are all of higher animals but the theory applies equally to invertebrates.

Feelings are the 'final causes' of actions. The term is a little misleading but I assume it refers to a true cause which precedes its effect. In this book it means the outcome of the last response to the same situation – the definition has to be revised somewhat when we consider generalization. I believe a teleological explanation of behaviour is scientifically valid but needs to be supplemented with an account of *how* learning takes place, and this is my main purpose.

Conditioning is seen as a vital adjunct to the learning process that can also be given both a teleological and analytical interpretation. Through it organisms achieve a degree of mastery over their surroundings by forecasting the more long-term results of their actions.

Animals use the experience gained in one setting to decide what to do in others. This is the essence of intelligence; a sort of 'common sense' in the old meaning of the phrase. There is however a much simpler form of learning by trial and error that only becomes apparent when exactly the same conditions arise again and which does not call for any special acumen. Surprisingly both abilities may be shown to be completely automatic and described in mechanistic terms by various logical formulae. These are the fundamental laws of human and animal psychology. They may be matched by diagrams indicating possible ways in which the brain works.

I am indebted to Dr J.D. Kenyon for reading the manuscript and making valuable suggestions, and to Dr S.F. Walker and others for their helpful advice. Thanks are due to my family for their unfailing encouragement over the years. I must also mention the tremendous enthusiasm and support of Pamela Grainger, expressed in a long correspondence, which helped to make the writing of the book a pleasure.

Preface

Animals possess innate powers of generalization which make wide use of the knowledge they gain from experience. An animal finds that a certain response to a certain situation has a certain outcome. It believes that the same response to the same situation will have the same outcome – and acts accordingly. This is a case of what we shall call simple trial-and-error learning. But it may also believe a different response to a different situation will also have that outcome – and again acts accordingly. Here intelligence may be involved. The two responses are said to be linked and, if they are really likely to have the same end, to converge. Links between converging responses explain how an animal behaves sensibly in situations it may never have encountered before. Reward (or punishment) of a response to one situation has an effect equal to rewarding (or punishing) responses to many others.

Some examples will help to make the issue clear. Suppose a hungry chicken is placed in a coop in which there is a tray of corn. It will soon run across the floor to find the food. From the same starting-point it will make the same response by the laws of simple trial-and-error learning. Now suppose we put a small screen in front of the tray. The bird will surely run round the obstacle to reach the food. This would be an example of primitive intelligence. The belief that if it runs round the screen its action is likely to have the same result as before is inborn. But the nature of that result – whether it was rewarding or punishing – depends on experience. If it had found the tray empty in the first place, for instance, it would probably not run round the screen in the second.

The chicken expects to find food behind the screen even though it may never have seen this object before. It does not have to rely on experience gathered in exactly the same setting to act wisely. Moreover its response is highly original. Convergence is not a matter of responding in the same way to different situations. Such simple generalization would take place if the screen had been placed to the left or right of the tray and not immediately in front of it. The situations would then be different but the bird would not have to alter its response to reach the food. This is not what we mean by convergence, which depends on a more complex form of generalization. Fig.1, Fig.2 and Fig.3 should make this clear.

Linked responses are interchangeable. Let us imagine the chicken had been placed with the screen in the coop in the first place, as in Fig.2. It would probably soon run round the obstruction and presumably discover

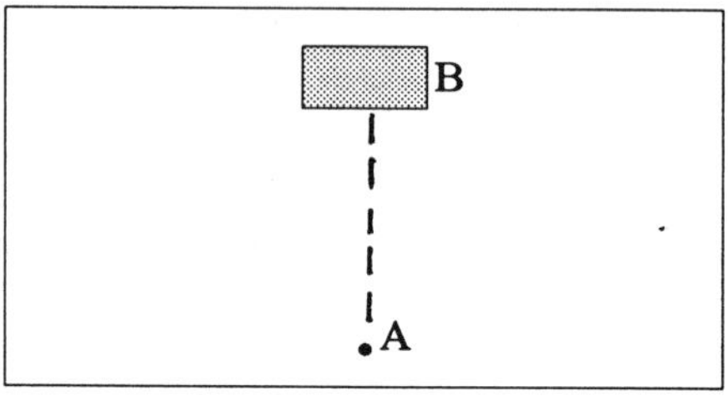

Fig.1. Simple trial and error. The chicken starts at A and runs straight to the tray B. Under exactly the same conditions it repeats the response.

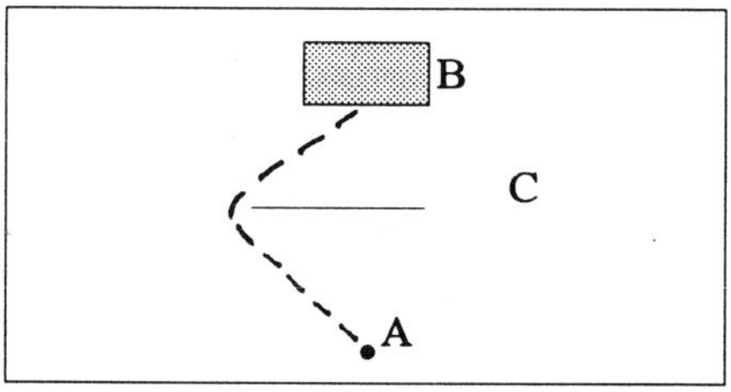

Fig.2. Convergence. The chicken starts at A and runs round the screen C to reach the tray B.

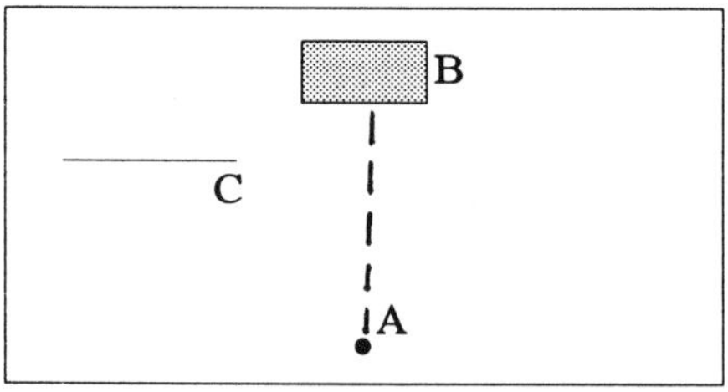

Fig.3. Simple generalization. The chicken starts at A and runs straight to the tray B. The situation differs from that in Fig.1 as the screen C is present.

the corn. This response would be repeated each time the same situation arose. If the screen had then been removed as in Fig.1 the bird would run straight to the tray. The rule is a general one. It always applies when situations eliciting linked responses are presented in reverse order.

Complex generalization may also be possible when the chicken approaches a goal from different starting-points, as in Fig.4 and Fig.5. The tray containing food is placed at E in the top left-hand corner of the coop. Suppose the chicken, starting at point D in Fig.4, runs to the tray E and finds it holds corn. If it is placed at D on further occasions it will continue to run to E, providing it always finds food there. Now suppose it is placed at point F, as in Fig.5. Its response may well be linked to the one in Fig.4 and bring it to the same goal E. If however the bird failed to discover food in the tray in the first instance it is doubtful whether it would run there in the second. So experience also plays a part. The two responses are interchangeable, as in the previous example.

In the wild convergence often takes place. A blackbird will fly home from numerous new directions in as many different ways. Sometimes its starting-points will be close together and its responses vary only slightly. On other occasions it may approach its nest from opposite ends and the responses linked will be quite dissimilar. It is moreover able to cope with changes in the appearance of its habitat which may be drastic. Trees sprout leaves, plants grow taller and hedges are clipped but the bird still finds its way to its brood. The daunting task of the animal psychologist is to explain such a mystery.

Much depends on what we mean by a situation. It is best thought of as a pattern of nervous impulses from groups of receptors in the animal's sense-organs. In the examples above we have been concerned mainly with visual images and this makes our work easier since human beings rely heavily on sight and can readily follow the reasoning. We should however appreciate other senses are involved in the definition of a situation and this is particularly true of animals. Henceforth our examples will be restricted to ones in which vision plays a dominant role purely to simplify the argument.

We must beware of thinking of situations too loosely. From the chicken's point of view those in Fig.4 and Fig.5 are quite different when defined in terms of patterns of light on its eyes. Similarly the three situations in Fig.1, Fig.2 and Fig.3 look very different to the bird, even though it sees them from the same starting-point. In fact the question of starting-points is not all that important. What matters are the patterns of impulses coming from the retina. If they are the same simple trial-and-

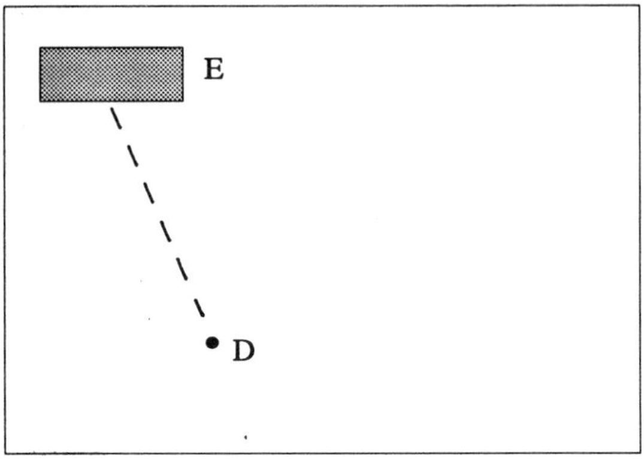

Fig. 4. Simple trial and error. The bird starts at D and learns to run to the tray E.

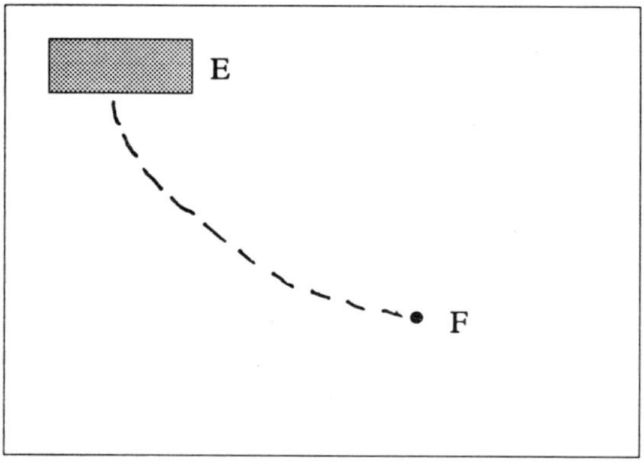

Fig. 5. Convergence. The chicken starts at F and runs to E.

error learning will take place. If different either simple or complex generalization may occur. The definition of a situation in this work is therefore subjective. In the text it is usually referred to as 'the cue'.

Convergence does not have to have a spatial connotation. It is incorrect to regard it always as a matter of moving from different starting-points to the same finishing-point, as in Fig.4 and Fig.5. In Fig.1 and Fig.2 the chicken began for example at the same point A. Equally, linked responses may carry a creature to different finishing-points. If it is likely the results of such responses are the same we are right to say they converge, regardless of where they bring the animal. Ideally rewards and punishments like situations are best conceived in terms of impulses fired from receptors and not as objects or events that happen in space at all.

Another example of convergence which may appear more complicated but is in principle the same as the previous ones is pictured in Fig.6 and Fig.7. The chicken starts at G in Fig.6 and runs to the tray H, where it finds corn. In Fig.7 the situation is then changed to one in which the bird begins at a different point I. The situation also differs from that in Fig.6 because the screen is placed at J. Neither of these changes prevent the bird from running straight to the tray H. Similarly in Fig.8 and Fig.9 the chicken first learns to run to the tray L from K and then runs round the screen N from a different starting-point M. Both the chicken's starting-points and surroundings have changed in these examples but complex generalization is probably still possible in each case. In both instances the principle of convergence applies as before: a different response to a different situation (as defined by the pattern of nervous impulses coming from the chicken's retina) leads to the same outcome.

Links are brought into being by evolution. They do not have to conform to any golden rule. The chicken in Fig.5 does not run in a straight line to the food though this would probably be the most sensible and efficient response. Even when actions are more reasonable and direct one will inevitably find some that are linked but do not converge, for animals are not perfect. In reality both the existence or absence of links may be adaptive or otherwise. Those between responses which fail to converge are likely to be eliminated by natural selection.

So far we have been concerned with feelings of satisfaction (of hunger) and dissatisfaction. But suppose when the chicken reached the tray in the first example it received a mild electric shock. The painful experience would have an effect similar to dissatisfaction. The bird would behave as if it had found the tray empty and in the second instance almost certainly not run round the screen. So we have three kinds of feeling or

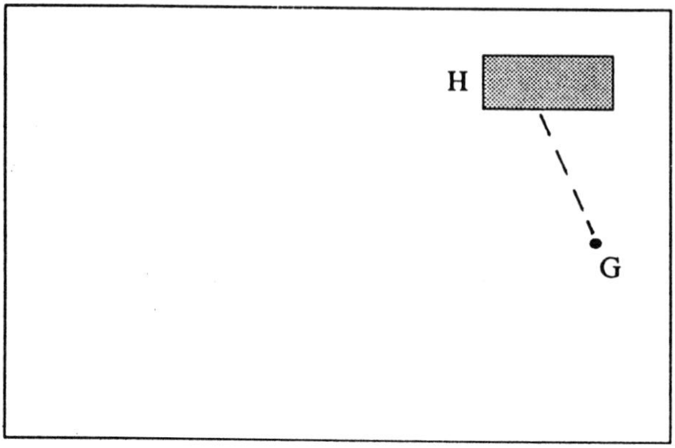

Fig.6. Simple trial and error. The chicken starts at G and learns to run to the tray H.

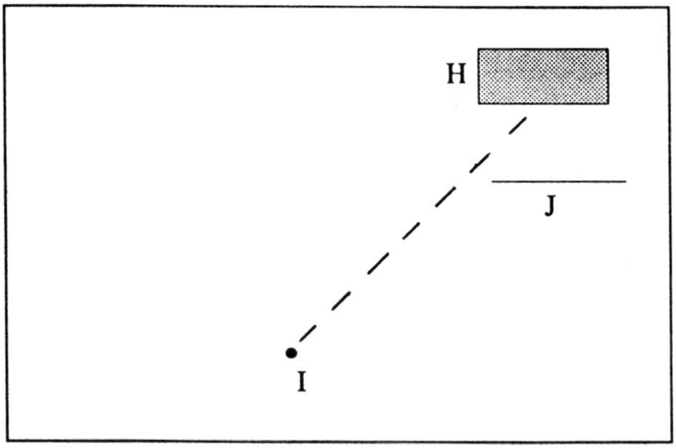

Fig.7. Convergence. The bird starts at I and runs straight to H. The screen at J does not interfere with the response.

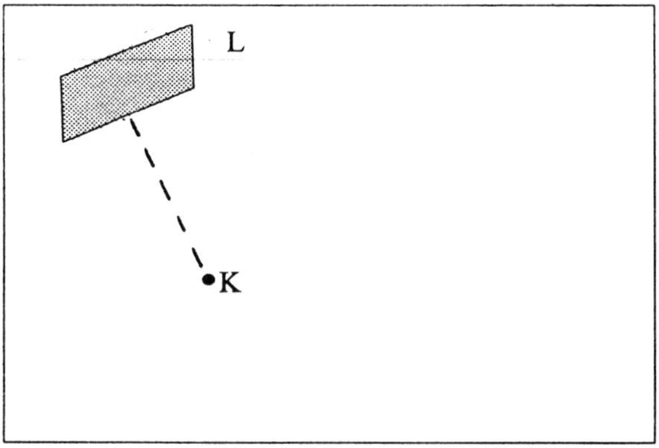

Fig.8. Simple trial and error. The chicken begins at K and runs straight to the tray of corn L. Subsequently when placed at K it repeats the response.

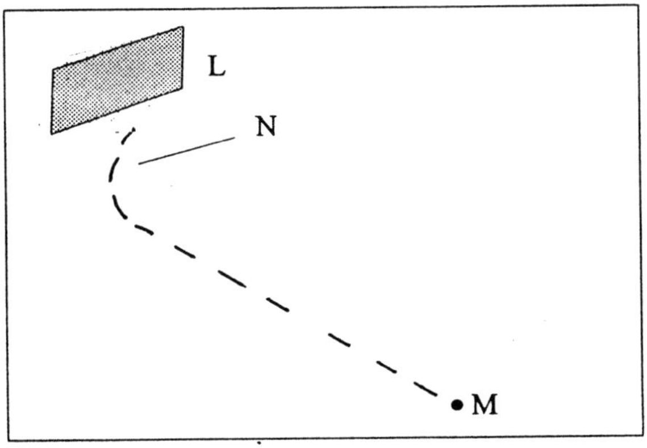

Fig.9. Convergence. When placed at a different starting-point M the bird runs round the screen N to reach the tray L.

sensation: satisfaction, which is rewarding, and dissatisfaction and pain, which are punishing.

Laws of problem-solving in animals were first laid down by E.L. Thorndike (*Animal Intelligence*, 1898). In a series of celebrated experiments he confined a cat in a puzzle-box. The animal had to unfasten a simple catch to open the door of the box. A piece of fish was placed outside to serve as an incentive. After making a large number of unsuccessful bids to escape the cat was seen to operate the catch, apparently by accident at first but in later trials more deliberately. Eventually it walked straight to the catch and opened the door without hesitation. Thorndike suggested that responses leading to satisfaction tended to become more strongly attached to the situation which evoked them, while in those resulting in dissatisfaction the tie was weakened, a principle which came to be called the *Law of Effect*.

His explanation is confusing, to say the least. It is better to visualize the box as containing not one or even a few but a huge variety of situations. When a response to one of them meets with satisfaction the effect is as if other responses to related situations had ended with that feeling too. The same rule of convergence applies to dissatisfaction. We shall see later how this apparently more complicated account of the animal's behaviour leads to a much simpler 'all-or-nothing' formulation of the above law, and one that may readily be translated in mechanical terms.

Conditioning also plays a part in learning. It consists of the taking on of rewarding or punishing power by neutral stimuli. A neutral stimulus which is followed shortly by the satisfaction of a need may be conditioned to act as a reward in its turn, while one followed by pain may become equally effective as a punishment. Second-order conditioning occurs when a neutral stimulus is followed shortly by one that has itself acquired rewarding or punishing power. Third and higher-order conditioning is also possible.

The learning of mazes by rats may to some extent be explained by the law of effect and conditioning. With this in mind C.L. Hull discovered blind-alleys were abandoned more rapidly near the end of a maze than at its beginning (*Principles of Behaviour*, 1943). Yet Hull like Thorndike never fully allowed for generalization. A rat seldom runs exactly the same path in successive trials, if only because it hardly ever begins at precisely the same point. Even slight changes in its circumstances require original movements, so simple trial-and-error learning is rare. Most animals are surprisingly versatile and their actions tend to become stereotyped only in the final stages of completing a task.

The rather difficult notion of avoidance conditioning, in which neutral stimuli acquire punishing power, should not be overlooked. Just as chains of responses leading to a reward may be gradually welded together by a combination of higher-order conditioning and trial-and-error learning so those ending in pain are liable to be broken up by the same processes. But in both cases it is well to remember generalization is always taking place. Goals may be approached through many different avenues and, conversely, an animal may be constrained to take avoiding action in response to an entirely unfamiliar setting.

A classical experiment is often quoted to illustrate the great flexibility of animal behaviour. Suppose a rat is trained to run a maze until its performance is faultless. It has been found that if the maze is half-submerged in water the creature will swim to the goal without difficulty. Here again a novel response (swimming) is needed to an altered situation (the water-filled maze). Information obtained in the first part of the experiment becomes useful in the second. Suppose, for example, the first choice-point in the maze is a T-junction, where the animal can turn either left or right. The response of swimming to the left would be likely to be linked to that of running to the left, while swimming to the right would probably be linked to running to the right. If the rat finds that running to the left is ultimately rewarded, while running in the opposite direction is not, then it will swim to the left when the occasion demands and so reach the goal, similar transfer of training taking place at all the other choice-points.

The experiment shows how different modes of locomotion may be used where appropriate. In the wild rats and many other animals need to know their way through elaborate systems of tunnels and these may frequently be filled with water. The survival-value of links here is obvious. This is not true of another experiment which nevertheless demonstrates the extraordinary ability of animals to solve problems set by man. After a rat has learned to run to the goal-box in a maze without error the whole apparatus is turned upside down. The rodent will then run to the same goal in the inverted maze, substituting right-hand turns for left-hand ones in its original position, and *vice versa*. The usefulness of such links can perhaps be explained as follows. Inquisitive rats in the wild must often explore the interiors of objects that may later be displaced or even completely upturned. Rats live in close proximity to man and these objects will often lie around, in rubbish dumps for example. When the animals find food it is an advantage if they know how to reach it when their environment is disturbed.

The foregoing examples are fairly straightforward ones taken from early studies in animal psychology. Yet there is no limit to the range of intelligence as we have defined it. As a rule the greater the dissimilarity between converging responses (and the situations which prompt them) the more this quality is thought to be in evidence. From a theoretical standpoint, however, the degree of difference is unimportant. The mechanism is the same whether it is great or small, as we shall see.

Introduction

This is an account of the effects of punishment and reward on behaviour and how they come about. It is based on what may be called the 'carrot and stick' theory of learning. Consider first the part played by the 'stick'. Suppose an animal (or human being) responds to a well-defined situation by making a move which results in punishment in the form of pain. The next time the situation arises the animal will make a different move.

The role of the 'carrot' is equally simple. Suppose the animal responds by making a move which is rewarded through the satisfaction of its needs. The next time the situation arises it will make the same move.

The 'carrot and stick' theory tells us *why* an animal makes a certain response to a situation. It explains behaviour in terms of so-called final causes. But we still want to know *how* an experience in the past determines an animal's action, particularly when the interval between cause and effect is indefinite. A convincing answer to this question has never been given.

The present work is intended to repair the omission. To that end the processes of nervous excitation and inhibition are shown to be subject to the rules of a two-valued logical calculus. Its algebra is developed at length in Chapter I. The system which emerges is a restricted but adequate one of only three undefined operations, depending on fundamental nervous mechanisms. The only other neurophysiological function postulated is the after-effect of either a stimulus or trace, which follows it after a short, fixed space of time. We are able to enlarge upon a plain teleology and show how the brain works in terms of these four functions.

Chapters II, III and IV are devoted to the analysis of simple trial-and-error learning, while Chapter V deals in part with intelligence viewed as an additional dimension. In Chapters VI and VII theorems worked out earlier are used to produce diagrams of nerve circuits in the brain.

Several papers at the end of the book were completed after it was written and show how the abstract treatment meted out so successfully there is highly effective in solving difficult problems of motivation. It would be impossible to do this without symbolic logic just as physics would be impossible without making extensive use of mathematical equations.

Learning through generalization and experience is a theme going back to Plato and Aristotle. Plato's theory of forms sought to explain general stimuli by the famous allegory of the cave, while Aristotle was the first to introduce the idea of final causes in animal and human conduct.

Chapter I

SYMBOLIC LOGIC

Relations between Stimuli

Stimuli are represented in this essay by the capital letter S followed by a small number or letter. For example, critical stimuli called elements are represented thus: S_1, S_2, S_3, etc. Like other stimuli elements are variables which are either present or absent. The value of presence is represented by the digit 1 and the value of absence by the digit 0. These digits must *not* be thought of as numbers.

The algebra of our two-valued logic consists of propositions stating that relations hold good between stimuli. Three of these relations are considered here: equivalence, contrariety and entailment. A pair of stimuli are said to be equivalent if when one is present the other is also present, while if one is absent the other is also absent. The table below, showing all four possible permutations of the two values of presence or absence which can be assigned to the elements S_1 and S_2, describes the relation.

		Equivalence
S_1	S_2	$S_1 \equiv S_2$
1	1	True
0	1	False
1	0	False
0	0	True

The relation of contrariety holds good between a pair of stimuli if it is impossible for both to be present at once. This relation is described again perfectly by a table. The sign of contrariety is a vertical stroke.

Contrariety

S_1	S_2	$S_1 \mid S_2$
1	1	False
0	1	True
1	0	True
0	0	True

The relation of entailment holds good between a pair of stimuli under the following conditions. If the first stimulus is absent the second may be either present or absent, but if the first stimulus is present the second must also be present, otherwise the relation does not hold good. Unlike equivalence and contrariety the order of the two stimuli in entailment is crucial.

Once again, the relation is described best by a table of values. The sign of entailment is $\supset$.

Entailment

S_1	S_2	$S_1 \supset S_2$
1	1	True
0	1	True
1	0	False
0	0	True

Relations may be reflexive, symmetrical and transitive. The relation of equivalence is reflexive, meaning that a stimulus is equivalent to itself.

$$S_1 \equiv S_1$$

A table tests the validity of the theorem above. It is true whichever value is assigned to the variable.

S_1	S_1	$S_1 \equiv S_1$
1	1	True
0	0	True

The relation of entailment is also reflexive, meaning that a stimulus entails itself.

$$S_1 \supset S_1$$

The validity of the theorem above can again be tested by means of a table.

S_1	S_1	$S_1 \supset S_1$
1	1	True
0	0	True

But the relation of contrariety is not reflexive, as is evident in the next table.

S_1	S_1	$S_1 \mid S_1$
1	1	False
0	0	True

Both equivalence and contrariety are symmetrical relations. The law of symmetry in the case of equivalence is stated as follows.

If, and only if, $S_1 \equiv S_2$ then $S_2 \equiv S_1$.

The symmetry of equivalence can be tested as follows. The truth-values in the columns are the same.

S_1	S_2	$S_1 \equiv S_2$	$S_2 \equiv S_1$
1	1	True	True
0	1	False	False
1	0	False	False
0	0	True	True

The law of symmetry in the case of contrariety is expressed thus. If, and only if, $S_1 \mid S_2$ then $S_2 \mid S_1$.

The table below tests the validity of the law above. Once again, we see that the truth-values in the columns are the same.

S_1	S_2	$S_1 \mid S_2$	$S_2 \mid S_1$
1	1	False	False
0	1	True	True
1	0	True	True
0	0	True	True

But entailment is not a symmetrical relation. The truth-values in the two columns below do not match.

S_1	S_2	$S_1 \supset S_2$	$S_2 \supset S_1$
1	1	True	True
0	1	True	False
1	0	False	True
0	0	True	True

The choice of symmetrical signs for equivalence and contrariety, and a non-symmetrical sign for entailment, is based on consideration of the tables above.

The relations of equivalence and entailment, unlike that of contrariety, are transitive.

If, $S_1 \equiv S_2$ and $S_2 \equiv S_3$ then $S_1 \equiv S_3$

If $S_1 \supset S_2$ and $S_2 \supset S_3$ then $S_1 \supset S_3$

We may read the laws above as follows. If S_1 is equivalent to S_2 and S_2 is equivalent to S_3 then S_1 is equivalent to S_3. Equally, if S_1 entails S_2 and S_2 entails S_3 then S_1 entails S_3. It is not true, however, that if S_1 is contrary to S_2 and S_2 is contrary to S_3 then S_1 is by necessity contrary to S_3. Tables could be constructed which verify these laws.

Summing up this section, equivalence is reflexive, symmetrical and transitive, contrariety is non-reflexive, symmetrical and non-transitive, and entailment is reflexive, non-symmetrical and transitive. The reflexivity of a relation involves one variable, the symmetry two and the transitivity three.

The table which follows lists these conclusions.

	Reflexivity	Symmetry	Transitivity
Equivalence	✓	✓	✓
Contrariety	x	✓	x
Entailment	✓	x	✓

We shall have occasion in the rest of this chapter to refer back to these propositions and tables frequently, in order to check the validity of theorems.

Logical Equivalence. Configurational Stimuli

The existence of a logical relation between stimuli depends on the logical nervous operations by which they are formed. An operation is represented by a pair of symbols denoting stimuli joined by a logical sign, the whole being enclosed in parentheses. The expression of a configurational stimulus drawn up in this way may itself be used as one of a pair to form the expression of a more complex stimulus, providing that once again parentheses are used to enclose the whole. In this manner configurational stimuli of any degree of complexity may be represented.

Examples of configurational stimuli are ($S_1 \ddagger S_2$) and ($S_1 . S_2$). They are evaluated in the tables below. Despite its name an element like S_1 or S_2 is itself configurational and may be expressed in terms of even simpler stimuli but we shall not do this here.

S_1	S_2	Variation ($S_1 \ddagger S_2$)
1	1	0
0	1	1
1	0	1
0	0	0

S_1	S_2	Conjunction ($S_1 . S_2$)
1	1	1
0	1	0
1	0	0
0	0	0

The value of presence or absence of a configurational stimulus is determined by the values assigned to its elements, so that in the case of variation in the table above the configurational stimulus ($S_1 \ddagger S_2$) is present only when one element is present and the other absent. When both elements are present or both absent the stimulus ($S_1 \ddagger S_2$) is absent. The sign '$\ddagger$' has the meaning of the connective 'or' when the possibility of both elements being present is ruled out. Its value is opposite to that of the relation of equivalence described in the first section and its sign is consequently a cancellation of the sign of that relation. But the use of the positive term 'variation' is preferred to 'non-equivalence' in this context. It is chosen to indicate a change in the value of the whole expression from that of the first element when the second element is present.

Similarly, in the case of conjunction in the table above the configurational stimulus ($S_1 . S_2$) is present only when both elements are present. If only one or neither are present the stimulus $(S_1 . S_2)$ is absent. The point '.' has the meaning of the connective 'and'. Its value is opposite to that of the relation of contrariety and it is accordingly cancelled by a vertical stroke to form the sign of that relation.

A third operation, called disjunction, is represented thus: $(S_1 \vee S_2)$. Disjunction may be defined in terms of variation and conjunction. Equally, conjunction may be defined in terms of variation and disjunction, but variation cannot be defined in terms of conjunction and disjunction. However, as we shall see, all three operations have similar properties and each one probably depends on a single nervous mechanism.

Disjunction is evaluated as follows.

Disjunction

S_1	S_2	$(S_1 \vee S_2)$
1	1	1
0	1	1
1	0	1
0	0	0

The configurational stimulus $(S_1 \vee S_2)$ is present when either one or other of the elements is present, or when both are present. The logical sign 'v', which is the first letter of the Latin word 'vel', has the meaning of the expression 'and/or'. This operation should be distinguished from variation, where the meaning of 'or' is exclusive.

A fourth operation, called occultation and represented by the expression $(S_1 \not\supset S_2)$, is defined in terms of variation and conjunction. Once again the use of a positive term is preferred to that of non-implication or non-entailment, for example. The letters Def. on the right of the equivalence below indicate that it is a definition. Another definition of occultation in terms of variation and conjunction, and a third in terms of variation and disjunction, are given later.

$$(S_1 \not\supset S_2) \equiv (S_1 \not\equiv (S_1 . S_2)) \quad \text{Def.(1)}$$

We can test the logical truth of the definition above stringently by constructing tables. This method of matrices, as it is called, should be distinguished from logical proof. Nevertheless it becomes a powerful tool for testing the validity of theorems. It is used extensively in this work.

S_1	S_2	$(S_1 . S_2)$	$(S_1 \not\equiv (S_1 . S_2))$	Occultation $(S_1 \not\supset S_2)$
1	1	1	0	0
0	1	0	0	0
1	0	0	1	1
0	0	0	0	0

A second table completes the verification. Reference should be made to the table of equivalence at the beginning of the chapter to check the truth-values in the fifth column.

S_1	S_2	$(S_1 \not\supset S_2)$	$(S_1 \not\equiv (S_1 . S_2))$	$(S_1 \not\supset S_2) \equiv (S_1 \not\equiv (S_1 . S_2))$
1	1	0	0	True
0	1	0	0	True
1	0	1	1	True
0	0	0	0	True

The sign of occultation '$\not\supset$' has the meaning in language of the expression 'but not'. Occultation is opposite in value to the relation of entailment and its sign is a cancellation of the sign of that relation.

A fifth function, the converse of occultation, may be defined in terms of that operation, as follows.

$$(S_1 \not\subset S_2) \equiv (S_2 \not\supset S_1) \text{ Def.}$$

A table confirming the logical truth of the definition above follows.

S_1	S_2	$(S_1 \not\subset S_2)$	$(S_2 \not\supset S_1)$	$(S_1 \not\subset S_2) \equiv (S_2 \not\supset S_1)$
1	1	0	0	True
0	1	1	1	True
1	0	0	0	True
0	0	0	0	True

Logic employs signs to represent relations and operations, variables and constants, as in mathematical algebra. But variables in mathematics can be assigned values of quantities which are infinite in number. In the algebra of logic on the other hand only two values can be assigned to the variables and there are accordingly only two constants corresponding to these values.

Of these two constants only one can be defined in terms of the operations described or defined above. This is the constant of stimulus-absence, represented by the digit 0. A definition in terms of variation follows.

$$0 \equiv (S_1 \not\equiv S_1) \text{ Def.}$$

Once again, the definition can be tested by means of a table.

S_1	S_1	$(S_1 \not\equiv S_1)$	0	$0 \equiv (S_1 \not\equiv S_1)$
1	1	0	0	True
0	0	0	0	True

The expression $(S_1 \not\equiv S_1)$ or its equivalent is described best as a configuration rather than a stimulus, as its value is fixed.

A complete table of values can now be drawn up. It consists of eight columns, which include the two elements, the five configurational stimuli and the constant of stimulus-absence.

S_1	S_2	$(S_1 \nequiv S_2)$	$(S_1 . S_2)$	$(S_1 \vee S_2)$	$(S_1 \not\supset S_2)$	$(S_1 \not\subset S_2)$	0
1	1	0	1	1	0	0	0
0	1	1	0	1	0	1	0
1	0	1	0	1	1	0	0
0	0	0	0	0	0	0	0.

The table above shows the limits of the system of operations on stimuli examined in this section. Disjunction can be defined in terms of variation and conjunction, or conjunction in terms of variation and disjunction. Equally, occultation and its converse can be defined in terms of variation and either conjunction or disjunction. But no other operations can be defined within the system. The logic of configurational stimuli is therefore a restricted one.

We see this in the bottom row of the table above, where the values are all absence. The restriction to five signs seems reasonable in this respect, that no configurational stimulus is present when all its elements are absent.

The details of the argument above are summed up in the table below.

Operation	Sign	Meaning	Opposite Relation	Sign
Variation	($\nequiv$)	'or'	Equivalence	$\equiv$
Conjunction	(.)	'and'	Contrariety	$\mid$
Disjunction	(v)	'and/or'		
Occultation	($\not\supset$)	'but not'	Entailment	$\supset$

The reader will see that no relation which is opposite to the operation of disjunction is listed in the table. The reason is that in the system under consideration this relation, which would normally be represented by the symbol $\downarrow$, cannot hold good between any pair of stimuli. The only logical theorems in which $\downarrow$ would appear as the main term are those in which it joins configurations that are both equivalent to the constant of stimulus-absence 0. Apart from this it has no application.

Logical Equivalence. Commutativity

The operations of variation, conjunction and disjunction are commutative, as is shown in the next three theorems.

$$(S_1 \not\equiv S_2) \quad \equiv \quad (S_2 \not\equiv S_1)$$

$$(S_1 \,.\, S_2) \quad \equiv \quad (S_2 \,.\, S_1)$$

$$(S_1 \vee S_2) \quad \equiv \quad (S_2 \vee S_1)$$

The laws of commutativity above are drawn up on the principle that the expressions on the right are reflections or mirror-images of those on the left. (Strictly speaking, this principle would be demonstrated better if symmetrical symbols were used for the variables as well as the signs.)

The commutative laws may be verified by the method of matrices. For example, the first theorem above is tested as follows.

S_1	S_2	$(S_1 \not\equiv S_2)$	$(S_2 \not\equiv S_1)$	$(S_1 \not\equiv S_2) \equiv (S_2 \not\equiv S_1)$
1	1	0	0	True
0	1	1	1	True
1	0	1	1	True
0	0	0	0	True

The law holds good whichever values of presence or absence are assigned to the variables. But the operation of occultation and its converse are not commutative. This is clear on studying the next table.

S_1	S_2	$(S_1 \not\supset S_2)$	$(S_2 \not\supset S_1)$	$(S_1 \not\supset S_2) \equiv (S_2 \not\supset S_1)$
1	1	0	0	True
0	1	0	1	False
1	0	1	0	False
0	0	0	0	True

The choice of non-symmetrical signs for occultation and its converse is based on these considerations.

Logical Equivalence. Associativity

Consider configurational stimuli with the patterns of parentheses (( )) and (( )). Six laws embodying these patterns can be drawn up in the restricted system of five signs, as follows.

$((S_1 \not\equiv S_2) \not\equiv S_3)$	$\equiv$	$(S_1 \not\equiv (S_2 \not\equiv S_3))$
$((S_1 \,.\, S_2) \,.\, S_3)$	$\equiv$	$(S_1 \,.\, (S_2 \,.\, S_3))$
$((S_1 \vee S_2) \vee S_3)$	$\equiv$	$(S_1 \vee (S_2 \vee S_3))$
$((S_1 \not\subset S_2) \not\supset S_3)$	$\equiv$	$(S_1 \not\subset (S_2 \not\supset S_3))$
$((S_1 \,.\, S_2) \not\supset S_3)$	$\equiv$	$(S_1 \,.\, (S_2 \not\supset S_3))$
$((S_1 \not\supset S_2) \not\supset S_3)$	$\equiv$	$(S_1 \not\supset (S_2 \vee S_3))$

The laws above, with the exception of the last, are drawn up in accordance with the following principle. The logical signs and variables (elements) on the left are reproduced exactly on the right, while the pattern of parentheses is reproduced by inversion. In other words, the pattern of parentheses on the right is a mirror-image of that on the left.

All six laws may be tested stringently by the method of matrices. We shall do this for the first, to show how the method is applied to theorems in which three variables are involved. There are eight possible permutations of the two values of these three variables, and these are listed in the three columns on the left of the table below.

S_1	S_2	S_3	$(S_1 \not\equiv S_2)$	$((S_1 \not\equiv S_2) \not\equiv S_3)$	$(S_2 \not\equiv S_3)$	$(S_1 \not\equiv (S_2 \not\equiv S_3))$
1	1	1	0	1	0	1
0	1	1	1	0	0	0
1	0	1	1	0	1	0
0	0	1	0	1	1	1
1	1	0	0	0	1	0
0	1	0	1	1	1	1
1	0	0	1	1	0	1
0	0	0	0	0	0	0

A second table completes the verification.

S_1	S_2	S_3	$((S_1 \not\equiv S_2) \not\equiv S_3)$	$(S_1 \not\equiv (S_2 \not\equiv S_3))$	$((S_1 \not\equiv S_2) \not\equiv S_3) \equiv (S_1 \not\equiv (S_2 \not\equiv S_3))$
1	1	1	1	1	True
0	1	1	0	0	True
1	0	1	0	0	True
0	0	1	1	1	True
1	1	0	0	0	True
0	1	0	1	1	True
1	0	0	1	1	True
0	0	0	0	0	True

Tables could be constructed which demonstrate the logical truth of the other five laws.

The first three laws above, in which the operations of variation, conjunction and disjunction figure, are called laws of associativity. The associative operations are also commutative, as we have seen. Using the laws of commutativity and associativity we may shift the position of the variables within the parentheses into any order we please.

We can omit the inner parentheses in the expressions without ambiguity, a practice which is adopted in this treatise. The theorems which follow define the convention.

$$(S_1 \not\equiv S_2 \not\equiv S_3) \quad \equiv \quad ((S_1 \not\equiv S_2) \not\equiv S_3) \quad \text{Def.}$$

$$(S_1 \,.\, S_2 \,.\, S_3) \quad \equiv \quad ((S_1 \,.\, S_2) \,.\, S_3) \quad \text{Def.}$$

$$(S_1 \vee S_2 \vee S_3) \quad \equiv \quad ((S_1 \vee S_2) \vee S_3) \quad \text{Def.}$$

Associations of any length can be drawn up with only the two outer parentheses, in line with the definitions above. Besides associations of four elements or more we sometimes speak of associations of two or even only one element.

All this formulation of the properties of logical operations is perhaps difficult until we recall that variation has the meaning of an exclusive 'or', conjunction of 'and' and disjunction of 'and/or'. It is taken for granted that these connectives are used without complicated punctuation in everyday language.

Logical Equivalence. Operators

It is possible to prove some important theorems from the six laws above. After proving each theorem it can be tested by the method of matrices. We begin by proving the theorem below as an example.

$$((S_1 \not\equiv S_2) \not\equiv S_3) \quad \equiv \quad ((S_1 \not\equiv S_3) \not\equiv S_2)$$

First, the appropriate associative law is reproduced below.

$$((S_1 \not\equiv S_2) \not\equiv S_3) \quad \equiv \quad (S_1 \not\equiv (S_2 \not\equiv S_3))$$

Using the appropriate commutative law the variables are shifted thus:

$$((S_2 \not\equiv S_1) \not\equiv S_3) \quad \equiv \quad ((S_2 \not\equiv S_3) \not\equiv S_1)$$

Now S_2 is replaced by S_1 and S_1 by S_2.

$$((S_1 \not\equiv S_2) \not\equiv S_3) \quad \equiv \quad ((S_1 \not\equiv S_3) \not\equiv S_2) \quad \text{QED}$$

Similarly, the theorems which follow can all be proved from the laws of commutativity and associativity and related theorems given above.

$$((S_1 \vee S_2) \vee S_3) \quad \equiv \quad ((S_1 \vee S_3) \vee S_2)$$

$$((S_1 \, . \, S_2) \, . \, S_3) \quad \equiv \quad ((S_1 \, . \, S_3) \, . \, S_2)$$

$$((S_1 \not\supset S_2) \not\supset S_3) \quad \equiv \quad ((S_1 \not\supset S_3) \not\supset S_2)$$

$$((S_1 \, . \, S_2) \not\supset S_3) \quad \equiv \quad ((S_1 \not\supset S_3) \, . \, S_2)$$

In the expressions on the left of the theorems above the formal arrangement consists of an element followed by two operators, each one consisting of a logical sign followed by an element. In the expressions on the right the position of the operators is switched, much in the same way that the position of the two variables is switched in the laws of commutativity above. The element S_1 is called the operand and remains in the same place on both sides of the theorems.

There are in fact four different types of operators in the system of logic examined in this work, corresponding to the first four signs. Examples are listed as follows: $\not\equiv S_2)$, $\vee S_2)$, $. \, S_2)$ and $\not\supset S_2)$.

In the last theorem above the operators differ in sign. We see that they can be switched as in the other theorems. The concept of an operator is derived from this theorem, as in the case of the other four theorems the same result is achieved if only the variable is switched. The need to switch both the variable and the logical sign which precedes it is made clear in the last theorem above.

The operation of conjunction is the basis of the nervous process of excitation, while the operation of occultation is the basis of inhibition. So the element S_1 in $(S_1 \, . \, S_2)$ is effective only when S_2 is present, while the same element S_1 in $(S_1 \not\supset S_2)$ is effective only when S_2 is absent. We speak of the excitatory influence of one stimulus on the action of another when its presence is necessary to implement that action, and an inhibitory influence of a stimulus on the action of another when its absence is necessary to implement it.

Logical Equivalence. Further Definitions

In the section on configurational stimuli a definition of occultation was given in terms of the commutative, associative operations of variation and conjunction, which are held to be primitive and based on fundamental nervous mechanisms. A second definition in terms of the same operations is given below.

$$((S_1 \not\supset S_2) \quad \equiv \quad (S_1 \,.\, (S_1 \not\equiv S_2)) \quad \text{Def. (2)}$$

The definition may be tested by the usual method of matrices.

S_1	S_2	$(S_1 \not\equiv S_2)$	$(S_1 \,.\, (S_1 \not\equiv S_2))$	$(S_1 \not\supset S_2)$
1	1	0	0	0
0	1	1	0	0
1	0	1	1	1
0	0	0	0	0

A third definition of occultation in terms of variation and disjunction is used in the next chapter to write definitions of cues. Disjunction is also a commutative, associative operation and is held to be primitive or based on a fundamental nervous mechanism.

$$((S_1 \not\supset S_2) \quad \equiv \quad ((S_1 \vee S_2) \not\equiv S_2) \quad \text{Def. (3)}$$

The definition above is tested thus:

S_1	S_2	$(S_1 \vee S_2)$	$((S_1 \vee S_2) \not\equiv S_2)$	$(S_1 \not\supset S_2)$
1	1	1	0	0
0	1	1	0	0
1	0	1	1	1
0	0	0	0	0

It will be seen that the third definition above contains the two operators v S_2) and $\not\equiv S_2$). When the operand is complicated, as in the case of the formation of cues, this definition is preferred.

The existence of the three definitions of occultation lends weight to the theory that inhibition is based on not one but two nervous mechanisms. The alternative view is that conjunction (excitation), occultation (inhibition) and disjunction are primitive or undefined operations and that variation should be defined in terms of occultation and disjunction, as follows.

$$(S_1 \not\equiv S_2) \quad \equiv \quad ((S_1 \not\supset S_2) \text{ v } (S_2 \not\supset S_1)) \text{ Def.}$$

The table below confirms the validity of this definition.

S_1	S_2	$(S_1 \not\supset S_2)$	$(S_2 \not\supset S_1)$	$((S_1 \not\supset S_2) \text{ v } (S_2 \not\supset S_1))$	$(S_1 \not\equiv S_2)$
1	1	0	0	0	0
0	1	0	1	1	1
1	0	1	0	1	1
0	0	0	0	0	0

A theorem similar in form to the definition above is used in the discovery of the intermittent trace in Chapter III.

Summing up so far, a number of theorems which are logically true, as the reader may find for himself by the method of matrices, are given with explanatory notes. A logical truth is one which remains true whichever values are assigned to the variables. In most cases the theorems are propositions which state that two configurational stimuli are equivalent. Some of these theorems are definitions. There are definitions of signs, of associations and of the constant of stimulus-absence. In the chapters which follow variables are defined in a contingent theory of learning, bearing in mind that a contingent truth does depend on the values assigned to its variables. The restricted logic examined above proves to be adequate for this purpose.

Logical Contrariety

So far we have been concerned mainly with theorems which state that the relation of equivalence holds good between certain stimuli. But there are other relations which are of interest in the theoretical study of the learning process. One of these is contrariety. Two stimuli are said to be contrary if it is impossible for both to be present at the same time.

An example of contrariety is the relation between a stimulus of need and its object. The action of a stimulus of need such as hunger is inhibited by the presence of its object, which is food. If we represent the stimulus of need by the expression $(S_1 \not\supset S_2)$ and its object by the variable S_2 then the presence of S_2 is contrary to the presence of $(S_1 \not\supset S_2)$. This is stated in the theorem which follows.

$$S_2 \mid (S_1 \not\supset S_2)$$

A table of values establishes the validity of the theorem above. The reader should refer to the table of contrariety given earlier to check the truth-values in the fourth column.

S_1	S_2	$(S_1 \not\supset S_2)$	$(S_2 \mid (S_1 \not\supset S_2)$
1	1	0	True
0	1	0	True
1	0	1	True
0	0	0	True

Cues, which are defined in Chapter II, are also contrary. For example, the first theorem of contrariety there is written in full as follows.

$$(S_1 \,.\, S_2 \,.\, S_3) \mid ((S_2 \,.\, S_3) \not\supset S_1)$$

Verification of the theorem above by the method of matrices is carried out in the tables below, referring to the table of contrariety at the beginning of the present chapter. Tables for all the other twenty theorems could be drawn up in the same way.

S_1	S_2	S_3	$(S_1 . S_2 . S_3)$	$(S_2 . S_3)$	$((S_2 . S_3) \not\supset S_1)$
1	1	1	1	1	0
0	1	1	0	1	1
1	0	1	0	0	0
0	0	1	0	0	0
1	1	0	0	0	0
0	1	0	0	0	0
1	0	0	0	0	0
0	0	0	0	0	0

The second table completes the verification.

S_1	S_2	S_3	$(S_1 . S_2 . S_3)$	$((S_2 . S_3) \not\supset S_1)$	$(S_1 . S_2 . S_3) \mid ((S_2 . S_3) \not\supset S_1)$
1	1	1	1	0	True
0	1	1	0	1	True
1	0	1	0	0	True
0	0	1	0	0	True
1	1	0	0	0	True
0	1	0	0	0	True
1	0	0	0	0	True
0	0	0	0	0	True

Logical Entailment

The third relation described at the beginning of the chapter is called entailment. A typical example is given below.

$$S_1 \supset (S_1 \text{ v } S_2)$$

A table establishes the validity of the theorem above. The reader should consult the table of entailment at the beginning of the chapter to check the truth-values in the fourth column.

S_1	S_2	$(S_1 \text{ v } S_2)$	$S_1 \supset (S_1 \text{ v } S_2)$
1	1	1	True
0	1	1	True
1	0	1	True
0	0	0	True

Entailment differs from equivalence in that the first stimulus in the relation may be absent while the second is present. It does not preclude the possibility that the second stimulus may be entailed by the presence of some other stimulus than the first. In the theorem above, for example, the second stimulus $(S_1 \text{ v } S_2)$ is entailed by S_2 as well as by S_1.

This is also evident in the next chapter where we consider the notion of a receptacle of disjunction of all the elements. The example given is $(S_1 \text{ v } S_2 \text{ v } S_3)$. It is entailed by either S_1, S_2 or S_3, or even by $(S_1 \text{ v } S_2)$, $(S_1 \text{ v } S_3)$ or $(S_2 \text{ v } S_3)$. The theorem below gives the first of these possibilities.

$$S_1 \supset (S_1 \text{ v } S_2 \text{ v } S_3)$$

The table below confirms the logical truth of this theorem.

S_1	S_2	S_3	$(S_1 \vee S_2 \vee S_3)$	$S_1 \supset (S_1 \vee S_2 \vee S_3)$
1	1	1	1	True
0	1	1	1	True
1	0	1	1	True
0	0	1	1	True
1	1	0	1	True
0	1	0	1	True
1	0	0	1	True
0	0	0	0	True

Entailment may also be found in the case of the disjunction called the varying complex, which is described in Chapter IV. Each stimulus of pain, need or fear of pain in the complex entails it. The presence of one or more of these stimuli brings about a change in future behaviour.

Chapter II

CUE AND RESPONSE

Definition of the Cue

In the 'carrot and stick' theory of learning described in the Introduction a key part is played by the situation to which the animal responds in the first place. This situation is a configurational stimulus which we call the cue. The cue is defined not only by the presence of elements but also by their absence. Bearing in mind the descriptions of associations in Chapter I, the elements which must be present to ensure the presence of the cue are called the positive association, which is a conjunction, while the elements which must be absent to ensure its presence are called the negative association, which is a disjunction. The positive and negative associations are joined by the operation of occultation, completing the definition.

Suppose the elements S_1, S_2 and S_3 comprise the positive association of a cue, while S_4, S_5 and S_6 comprise the negative association. The cue is then defined as $((S_1 \,.\, S_2 \,.\, S_3) \not\supset (S_4 \vee S_5 \vee S_6))$.

The assertion that it is present means that S_1 *and* S_2 *and* S_3 are present (representing a conjunction) but neither S_4 *nor* S_5 *nor* S_6 is present (representing a disjunction). The order of the elements in the associations is immaterial, as we have seen. They are linked by the sign of occultation, which has the meaning 'but not'.

For the sake of simplicity we are now going to consider the formation of cues by permutations of presence and absence of just the three elements S_1, S_2 and S_3. With this limited field we can still demonstrate many important laws of the learning process.

Cues are represented by the symbols Sa, Sb, Sc, Sd, Se, Sf, Sg and Sh. With N elements there are 2^N possible situations. The number of cues that can be defined in the system is one less than this number. So with three independent elements seven cues can be defined, as below.

S_1	S_2	S_3			
1	1	1	Sa	$\equiv$	$(S_1 \,.\, S_2 \,.\, S_3)$
0	1	1	Sb	$\equiv$	$((S_2 \,.\, S_3) \not\supset S_1)$
1	0	1	Sc	$\equiv$	$((S_1 \,.\, S_3) \not\supset S_2)$
0	0	1	Sd	$\equiv$	$(S_3 \not\supset (S_1 \vee S_2))$
1	1	0	Se	$\equiv$	$((S_1 \,.\, S_2) \not\supset S_3)$
0	1	0	Sf	$\equiv$	$(S_2 \not\supset (S_1 \vee S_3))$
1	0	0	Sg	$\equiv$	$(S_1 \not\supset (S_2 \vee S_3))$
0	0	0	Sh	$\equiv$	—

It was pointed out in Chapter I that associations of two or even only one element are possible. This is the case in the definitions above. The cue Sa consists of a positive association of three elements on its own. Sb, Sc and Se consist of positive associations of two elements with negative associations of one. Sd, Sf and Sg consist of positive associations of one element with negative associations of two.

The cue Sh which has no positive association cannot be defined within the system of configurational stimuli examined in Chapter I. This anomaly arises out of the restriction of the logic to five signs and one constant.

Another way in which Sd, Sf and Sg can be defined follows from a theorem also given in Chapter I. This is reproduced below.

$$((S_1 \not\supset S_2) \not\supset S_3) \equiv (S_1 \not\supset (S_2 \vee S_3))$$

The expression on the right corresponds to Sg. Similar definitions may be drawn up for the cues Sd and Sf.

The cues are listed below after making these changes.

S_1	S_2	S_3			
1	1	1	Sa	≡	$(S_1 . S_2 . S_3)$
0	1	1	Sb	≡	$((S_2 . S_3) \not\supset S_1)$
1	0	1	Sc	≡	$((S_1 . S_3) \not\supset S_2)$
0	0	1	Sd	≡	$((S_3 \not\supset S_1) \not\supset S_2)$
1	1	0	Se	≡	$((S_1 . S_2) \not\supset S_3)$
0	1	0	Sf	≡	$((S_2 \not\supset S_1) \not\supset S_3)$
1	0	0	Sg	≡	$((S_1 \not\supset S_2) \not\supset S_3)$
0	0	0	Sh	≡	—

A more systematic definition of the cue can be written by putting first the constant of stimulus-presence (the digit 1) followed by conjunctions of each member of the positive association and by occultations of each member of the negative association. But as explained above we cannot define the constant of presence in terms of the five signs of the system, only the constant of stimulus-absence (the digit 0). So it is only possible to approximate to the constant of presence by writing a disjunction of *all* the elements, which we call the receptacle. This is present when at least one element is present. The receptacle in the next eight definitions is $(S_1 \vee S_2 \vee S_3)$. There is an excitatory influence on the action of the receptacle by each member of the positive association and an inhibitory influence by each member of the negative association.

S_1	S_2	S_3			
1	1	1	Sa	≡	$((((S_1 \vee S_2 \vee S_3) . S_1) . S_2) . S_3)$
0	1	1	Sb	≡	$((((S_1 \vee S_2 \vee S_3) \not\supset S_1) . S_2) . S_3)$
1	0	1	Sc	≡	$((((S_1 \vee S_2 \vee S_3) . S_1) \not\supset S_2) . S_3)$
0	0	1	Sd	≡	$((((S_1 \vee S_2 \vee S_3) \not\supset S_1) \not\supset S_2) . S_3)$
1	1	0	Se	≡	$((((S_1 \vee S_2 \vee S_3) . S_1) . S_2) \not\supset S_3)$
0	1	0	Sf	≡	$((((S_1 \not\supset S_2 \vee S_3) \not\supset S_1) . S_2) \not\supset S_3)$
1	0	0	Sg	≡	$((((S_1 \vee S_2 \vee S_3) . S_1) \not\supset S_2) \not\supset S_3)$
0	0	0	0	≡	$((((S_1 \vee S_2 \vee S_3) \not\supset S_1) \not\supset S_2) \not\supset S_3)$

The value of the eighth expression is fixed as absence. It is a configuration rather than a cue. The response to the situation defined by the absence of all the elements must therefore be one of instinctive inaction; instinctive because it is unaffected by experience.

The definitions are written in accordance with the rules of the system. The unusual case when all the elements are absent become insignificant as the number of elements in the receptacle increases.

The definitions of cues in terms of the receptacle and conjunctions or occultations, which are given above, can also be expressed in terms of the commutative, associative operations of variation, conjunction and disjunction, since occultation can be defined in those terms. But the use of the first two definitions of occultation in terms of variation and conjunction gives definitions of cues of inordinate length. For this reason the definition of occultation in terms of variation and disjunction is preferred.

S_1 S_2 S_3								
1 1 1	Sa ≡	$((((S_1 \vee S_2 \vee S_3)$		$. S_1)$		$. S_2)$		$. S_3)$
0 1 1	Sb ≡	$(((((S_1 \vee S_2 \vee S_3)$	$\vee S_1)$	$\not\equiv S_1)$		$. S_2)$		$. S_3)$
1 0 1	Sc ≡	$(((((S_1 \vee S_2 \vee S_3)$		$. S_1)$	$\vee S_2)$	$\not\equiv S_2)$		$. S_3)$
0 0 1	Sd ≡	$((((((S_1 \vee S_2 \vee S_3)$	$\vee S_1)$	$\not\equiv S_1)$	$\vee S_2)$	$\not\equiv S_2)$		$. S_3)$
1 1 0	Se ≡	$(((((S_1 \vee S_2 \vee S_3)$		$. S_1)$		$. S_2)$	$\vee S_3)$	$\not\equiv S_3)$
0 1 0	Sf ≡	$(((((S_1 \vee S_2 \vee S_3)$	$\vee S_1)$	$\not\equiv S_1)$		$. S_2)$	$\vee S_3)$	$\not\equiv S_3)$
1 0 0	Sg ≡	$((((((S_1 \vee S_2 \vee S_3)$		$. S_1)$	$\vee S_2)$	$\not\equiv S_2)$	$\vee S_3)$	$\not\equiv S_3)$
0 0 0	0 ≡	$(((((((S_1 \vee S_2 \vee S_3)$	$\vee S_1)$	$\not\equiv S_1)$	$\vee S_2)$	$\not\equiv S_2)$	$\vee S_3)$	$\not\equiv S_3)$

Cues are contrary, as the table below shows.

S_1	S_2	S_3	Sa	Sb	Sc	Sd	Se	Sf	Sg
1	1	1	1	0	0	0	0	0	0
0	1	1	0	1	0	0	0	0	0
1	0	1	0	0	1	0	0	0	0
0	0	1	0	0	0	1	0	0	0
1	1	0	0	0	0	0	1	0	0
0	1	0	0	0	0	0	0	1	0
1	0	0	0	0	0	0	0	0	1
0	0	0	0	0	0	0	0	0	0

From the table above we can draw up theorems of contrariety. As it is not a transitive relation twenty-one theorems must be written to establish the complete contrariety of the seven cues, in the sense that only one can be present at once.

Sa	Sb
Sa	Sc
Sa	Sd
Sa	Se
Sa	Sf
Sa	Sg
Sb	Sc
Sb	Sd
Sb	Se
Sb	Sf
Sb	Sg
Sc	Sd
Sc	Se
Sc	Sf
Sc	Sg
Sd	Se
Sd	Sf
Sd	Sg
Se	Sf
Se	Sg
Sf	Sg

The first theorem of contrariety above is tested by the method of matrices in Chapter I. The other twenty theorems could be tested in the same way.

Now we have seen the number of cues is an exponential function of the number of elements. For the sake of simplicity a system of three elements and seven cues has been considered. The actual numbers of these stimuli are of course much greater. With ten elements there are over a thousand cues, with twenty over a million, with thirty over a thousand million and with forty over a million million. It does therefore seem probable that the number of elements is in *at least double figures but not more than double figures* if the number of cues is to be neither too small nor too great.

Compared with the cues, then, the elements are few in number and must in their turn be defined in terms of even simpler stimuli to give a complete picture of the environment. In the final analysis the only stimuli which are not configurational are those which discharge individual receptors at a microscopic level in the sense-organs. All the logical operations described above may be involved in the formation of elements from these numerous irreducible components.*

But the cues remain contrary whatever the logical composition of the elements. This follows from the laws of sensory interaction given above and constitutes a theory of perception in itself. Later we shall describe the motor interaction of regular and intermittent traces to form comprehensive traces, which are similar in their logical pattern to the cues.

The Comprehensive Trace

Whether or not a cue elicits a certain response depends on the presence or absence respectively of a variable called a comprehensive trace. When present it has an excitatory influence on the action of the cue, so that a response is made to the cue if the trace is present simultaneously.

---

*Sometimes this means that the elements are not independent variables. For example, let S_I and S_{II} be stimuli which discharge receptors in the sense-organs. Let us suppose that two of the elements are defined as follows.

$$S_2 \equiv S_I$$
$$S_3 \equiv (S_I \vee S_{II})$$

From these definitions it is evident that S_2 and S_3 are not independent, since the following theorem holds good.

$$S_2 \supset S_3$$

When this is true Se and Sf are both cues which are invariably absent. Each depends on the presence of S_2 and absence of S_3.

The comprehensive trace is the record of a decision of the animal to respond in a certain way to a certain cue. If on the last occasion the cue was presented a response was made which had no aversive outcome (such as pain or need) the comprehensive trace corresponding to that cue and response remains present, in a way that will be explained shortly. On the next occasion the cue is presented the same response will be elicited, by the conjunction of cue and comprehensive trace.

On the other hand, if on the last occasion the cue was presented a response was made which did indeed have an aversive outcome, the comprehensive trace corresponding to that cue and response becomes absent, and the comprehensive trace corresponding to the same cue but the next response in the animal's repertoire is made present. On the next occasion the cue is presented this new response will be elicited.

It is a basic tenet of the theory that only decisions are remembered. No record is kept of the experiences that prompt them.

The comprehensive trace is represented by a capital letter T followed firstly by the same small letter which denotes the appropriate cue, and secondly by a number in brackets indicating the action that will ensue when this trace and its cue are present together. Some examples are given as follows: Ta[1], Ta[5], Tb[5].

Stimuli of Action

One of the main aims of this work is to define the relation between behaviour and experience. The way we have set out to do this is to define the logical relation between stimuli which elicit responses and those which elicit feelings. The former are called stimuli of action. Each of them is represented by a capital letter S followed by a number in brackets corresponding to the response made, thus, S[0], S[1], S[2], etc.

In the previous section we saw that a cue becomes effective in instigating an action when the corresponding comprehensive trace is also present. The stimulus of action is defined as a disjunction of such effective cues. Hence, as we saw in Chapter I, it is entailed by them. The examples below illustrate this point. It is the rule that the number of the response made is the same on both sides of each theorem.

$$(Sa \;.\; Ta[1]) \quad \supset \quad S[1]$$

$$(Sa \;.\; Ta[5]) \quad \supset \quad S[5]$$

$$(Sb \;.\; Tb[5]) \quad \supset \quad S[5]$$

The theorems above show that the stimulus of action S[1] is entailed by the effective cue (Sa . Ta[1]) and the stimulus of action S[5] by either of the effective cues (Sa . Ta[5]) or (Sb . Tb[5]).

As we shall see later comprehensive traces of the same cue but different numbers of responses are contrary. So Ta[0] is contrary to Ta[1], Ta[1] is contrary to Ta[2] and Ta[0] is contrary to Ta[2], etc. Hence effective cues must also be contrary. For example, the effective cues (Sa . Ta[1]), (Sa . Ta[5]) and (Sb . Tb[5]) are completely contrary in the sense that only one of them can be present at once. (Sb . Tb[5]) is contrary to (Sa. Ta[1]) and to (Sa . Ta[5]) because Sb is contrary to Sa, while (Sa . Ta[1]) is contrary to (Sa . Ta[5]) because Ta[1] is contrary to Ta[5].

The contrariety of effective cues implies the contrariety of stimuli of action, as each of the former entails one and only one of the latter. So S[0] is contrary to S[1], S[1] is contrary to S[2] and S[0] is contrary to S[2], etc. In other words, only one response can be made at once. This is a central feature of the theory.

The contrariety of effective cues also implies that in the definition of each stimulus of action the sign of disjunction may be replaced throughout by that of variation. The sign of disjunction 'v' means 'and/or' but when the possibility that more than one stimulus is present is ruled out, as in the case of effective cues that are contrary, the sign of variation '$\ddagger$' meaning 'or' in its exclusive sense may be used instead. Either mechanism is equally good.

The response of doing nothing, which is called inaction, is made when all the elements are absent, a state of affairs which cannot be defined within the system of operations described in Chapter I. The occasions on which this phenomenon occurs decrease greatly as the number of elements in the receptacle of the cue increases.

The response of inaction must be distinguished from the first response to Sa in the animal's repertoire and therefore the first response in the life of the animal to this cue. The stimulus of action which elicits this response is represented by the symbol S[0]. We shall see later why this symbol has been chosen. For the present it is sufficient to say that S[0] is a variable like other stimuli and can take values of presence 1 or absence 0.

Finally, at a microscopic level again, an effector in the muscles is innervated by a stimulus which is equivalent to a disjunction of one or more stimuli of action. As the latter are contrary the association may equally be a variation. Once again, each stimulus of action entails the stimulus which innervates the effector and, since entailment is a transitive relation, each effective cue entails it too. Unlike the stimuli of action the stimuli which innervate effectors are not necessarily contrary.

Stages in the formation of stimuli are listed below. Those marked by the letter C. are contrary. Only one of them can be present at once.

(1) Stimuli which discharge receptors.	Sensory stimuli
(2) Elements. e.g. S_1	
(3) Cues. C. e.g. Sa	
(4) Effective cues. C. e.g. (Sa . Ta[0])	Motor stimuli
(5) Stimuli of action. C. e.g. S[0]	
(6) Stimuli which innervate effectors.	

Chapter III

THE MECHANISMS OF LEARNING

Behaviour and Experience Series

A behaviour series is a convenient way of symbolizing successive responses to a cue. The number of each response of action is written in brackets. When the cue under consideration is absent the space left in the series is filled by an asterisk. This indicates that some other cue elicits the response.

A typical behaviour series is written below. The cue which elicits the response is Sa.

	1	2	3	4	5	6	7	8	9	10
Sa elicits	*	[0]	*	*	[1]	[1]	*	[2]	*	[2]

In the series above the cue Sa elicits a response in spaces 2, 5, 6, 8 and 10. The responses go either in numerical order or are repeated. Thus, the response [0] in space 2 is followed by the response [1] in space 5. In space 6 this response is repeated. In space 8 the cue Sa elicits the next response [2] in the animal's repertoire, which is repeated in space 10.

An experience series gives the successive values of presence or absence of a stimulus of pain or some need such as hunger, represented by the symbol of a dotted capital $\dot{S}$. As before, the digit 1 represents presence and the digit 0 absence. An example is written as follows.

	1	2	3	4	5	6	7	8	9	10
$\dot{S}$ (pain or need)	0	0	1	0	0	0	1	1	0	0

Bringing the two simultaneous series together it is possible to show the effect of experience on behaviour during learning, as follows.

Example 1	1	2	3	4	5	6	7	8	9	10
Sa elicits	*	[0]	*	*	[1]	[1]	*	[2]	*	[2]
$\dot{S}$ (pain or need)	0	0	1	0	0	0	1	1	0	0

In the behaviour series above the first response to Sa in space 2 is [0]. This response has an aversive outcome (the presence of $\dot{S}$) in space 3 of the experience series. So in space 5 the cue Sa elicits the next response in the repertoire which is [1].

The response [1] in space 5 of the behaviour series has no aversive outcome in space 6 of the experience series. It is therefore repeated. But the outcome of this second response of [1] is aversive in space 7 of the experience series, and in consequence the next response in the repertoire, which is [2], is elicited by Sa in space 8.

The response [2] in space 8 of the behaviour series has no aversive outcome in space 9 of the experience series. This response is accordingly repeated in space 10.

Properties of these series can be demonstrated by further examples.

Example 2	10	11	12
Sa elicits	[4]	*	[4]
$\dot{S}$ (pain or need)	1	0	0

In the example above response [4] has no aversive outcome in space 11 of the experience series, and is repeated. The fact that response [4] and $\dot{S}$ coincide in space 10 does not affect the issue.

Example 3	10	11	12
Sa elicits	[4]	*	[5]
$\dot{S}$ (pain or need)	0	1	0

In this example response [4] is followed after one space by the presence of $\dot{S}$. In consequence, response [5] is elicited in space 12.

Example 4	8	9	10	11	12	13
Sa elicits	[5]	*	*	[5]	*	[5]
$\dot{S}$ (pain or need)	0	0	1	0	0	0

In Example 4 above response [5] in space 8 is not followed in space 9 by the presence of $\dot{S}$. Response [5] in this case is therefore repeated, even though $\dot{S}$ is present after two spaces in space 10. The aversive outcome must follow immediately.

Equally, the fact that $\dot{S}$ in space 10 precedes [5] in space 11 again does not affect the issue, and the cue Sa elicits the same response [5] in space 13. The stimulus of pain or need must follow and not precede the response if it is to change.

Example 5	20	21	22	23	24	24
Sa elicits	[7]	*	[0]	*	*	[1]
$\dot{S}$ (pain or need)	0	1	0	1	0	0

A repertoire of eight responses ends with the response of action [7], as the first response is [0]. When [7] is followed by a stimulus of pain or need, as in Example 5 above, the next presentation of the cue Sa prompts the response [0]. The repertoire thus begins anew.

Example 6	18	19	20	21
Sa elicits	*	[3]	[4]	*
$\dot{S}$ (pain or need)	0	0	1	0

In Example 6 above the response [3] when punished by the presence of $\dot{S}$ changes to [4] in the same space. The effect of $\dot{S}$ is immediate.

Example 7	6	7	8	9	10	11	12
Sa elicits	[3]	*	*	*	*	*	[3]
$\dot{S}$ (pain or need)	0	0	0	0	1	0	0

In Example 7 the decision to respond in a certain way to a cue is remembered over a longer period of time. The presence of $\dot{S}$ in space 10 does not affect the issue.

Example 8	12	13	14	15	16	17	18	19
Sa elicits	[5]	*	*	*	*	*	*	[6]
$\dot{S}$ (pain or need)	0	1	0	0	0	0	0	0

In Example 8 above the decision to respond to the cue Sa by the action [6] is made in space 13, and implemented in space 19, after an interval of seven spaces between successive presentations of the cue.

In the eight examples above we have shown why an animal makes a certain response to a cue. Its action is determined by the outcome of the last response to the same cue. This is what we mean by a teleological explanation of behaviour. But we also want to know how this selection of responses is brought about, in terms of operations in the nervous system. More particularly, we want to know how a decision to respond in a certain way to a cue is retained in the memory over an indefinite period of time.

The answer to these questions is to be found in the pages which follow. To simplify matters a ninth example of learning is provided which is summarized by a behaviour and experience series similar to those in the examples above. The logical operations underlying the learning process in this example are worked out in full.

The Regular Trace

The order of events in the learning process is as follows.

(1) The cue

(2) The response of the animal to the cue

(3) The reaction of the environment to the response

Stimuli function in two ways. They may figure in the cue (1) before the response, and they may also figure in the reaction (3) after. In this second role they are described as stimuli of reaction. Stimuli of reaction of pain or need form a disjunction which we call the varying complex. Stimuli which do not figure directly in it are called neutral.

The interval between the presentation of the cue (1) and the reaction of the environment (3) is called a space. The response is defined as the total activity of the animal during this time.

Each cue is followed after one space by an after-cue. The after-cue is denoted by the same symbol as the corresponding cue with a stress, thus, Sá, Sb́. If the cue is present before the response the after-cue will be present after the response. Conversely, if the cue is absent before the response the after-cue will be absent after.

In Chapter II the concept of a comprehensive trace was introduced. The comprehensive trace is formed by the interaction of a variable called the regular trace, represented by the symbols Ta1, Tb1, together with intermittent traces. These traces are all variables which like stimuli are either present or absent. They interact to form the comprehensive trace in exactly the same way that elements do to form cues. This will be made clear later.

A regular trace is followed after one space by an after-trace, which is denoted by the same symbol with a stress, thus, Tá1, Tb́1. If the regular trace is present before the response its after-trace will be present after. Conversely, if the regular trace is absent before the response the after-trace will be absent after.

Finally, the varying complex is represented by the dotted capital Ṡ. This symbol appeared in the previous section.

It is possible to draw up definitions of the regular traces with these symbols. Variation in these definitions is called primary.

$$Ta1 \equiv (T\acute{a}1 \nparallel (S\acute{a} \,.\, \dot{S}))$$

$$Tb1 \equiv (T\acute{b}1 \nparallel (S\acute{b} \,.\, \dot{S}))$$

In the definitions of the regular traces above it is evident that the stimuli of reaction (pain or need) which form the varying complex have an excitatory influence on the action of the after-cue.

The argument below shows that when the varying complex is absent after the response the value of the regular trace remains the same.

If	$Ta1 \equiv (T\acute{a}1 \nparallel (S\acute{a} \,.\, 0))$
then	$Ta1 \equiv (T\acute{a}1 \nparallel 0)$
and	$Ta1 \equiv T\acute{a}1$

Equally, when the cue is absent before the response the after-cue will be absent after and the trace will persist.

If	$Ta1 \equiv (T\acute{a}1 \nparallel (0 \,.\, \dot{S}))$
then	$Ta1 \equiv (T\acute{a}1 \nparallel 0)$
and	$Ta1 \equiv T\acute{a}1$

Persistence of the regular trace as above is the basis of memory. Its value remains the same between successive presentations of the cue.

The table below shows the conditions under which primary variation takes place.

Tá1	Sá	Ṡ	(Sá . Ṡ)	Ta1 (Tá1 ≢ (Sá . Ṡ))	Tá1		Ta1
1	1	1	1	0	1	varies to	0
0	1	1	1	1	0	varies to	1
1	0	1	0	1		no variation	
0	0	1	0	0		no variation	
1	1	0	0	1		no variation	
0	1	0	0	0		no variation	
1	0	0	0	1		no variation	
0	0	0	0	0		no variation	

Intermittent Traces

The definition of a regular trace is reproduced below.

Ta1 ≡ (Tá1 ≢ (Sá . Ṡ))

Changing both sides of the definition above in the same way produces the following.

(Tá1 ≢ Ta1) ≡ (Tá1 ≢ (Tá1 ≢ (Sá . Ṡ)))

Simplifying the right-hand side of the theorem above gives us the following theorem.

(Tá1 ≢ Ta1) ≡ (Sá . Ṡ)

Now primary variation takes place when the after-cue and the varying complex are present jointly. In other words, primary variation takes place when (Sá . Ṡ) is present. From the theorem above we conclude that primary variation takes place when (Tá1 ≢ Ta1) is present.

The expression (Tá1 ≢ Ta1) can be analysed as follows. The validity of this theorem can be checked by the method of matrices, as in Chapter I.

$$(T\acute{a}1 \not\equiv Ta1) \equiv ((T\acute{a}1 \mathrel{\text{þ}} Ta1) \vee (Ta1 \mathrel{\text{þ}} T\acute{a}1))$$

Intermittent variation every *second* stimulation by $\dot{S}$ takes place when just one of the two expressions $(T\acute{a}1 \mathrel{\text{þ}} Ta1)$ or $(Ta1 \mathrel{\text{þ}} T\acute{a}1)$ is present. Let us assume that the first expression $(T\acute{a}1 \mathrel{\text{þ}} Ta1)$ brings about intermittent variation.* We must introduce the concepts of an intermittent trace Ta2 and its after-effect Tá2 to write a definition, thus.

$$Ta2 \equiv (T\acute{a}2 \not\equiv (T\acute{a}1 \mathrel{\text{þ}} Ta1))$$

By exactly the same reasoning a definition for the intermittent trace Ta3, which registers variation every *fourth* stimulation by $\dot{S}$, is drawn up as follows.

$$Ta3 \equiv (T\acute{a}3 \not\equiv (T\acute{a}2 \mathrel{\text{þ}} Ta2))$$

Alternative definitions in terms of the commutative, associative operations of variation and conjunction can be proved from the definitions of the regular and intermittent traces above, thus.

$$Ta1 \equiv (T\acute{a}1 \not\equiv (S\acute{a} \,.\, \dot{S}))$$

$$Ta2 \equiv (T\acute{a}2 \not\equiv (S\acute{a} \,.\, \dot{S} \,.\, T\acute{a}1))$$

$$Ta3 \equiv (T\acute{a}3 \not\equiv (S\acute{a} \,.\, \dot{S} \,.\, T\acute{a}1 \,.\, T\acute{a}2))$$

To prove the second of these definitions we begin by reproducing the definition of the intermittent trace Ta2 above, as follows.

$$Ta2 \equiv (T\acute{a}2 \not\equiv (T\acute{a}1 \mathrel{\text{þ}} Ta1))$$

Replace the regular trace Ta1 by its definition $(T\acute{a}1 \not\equiv (S\acute{a} \,.\, \dot{S}))$

$$Ta2 \equiv (T\acute{a}2 \not\equiv (T\acute{a}1 \mathrel{\text{þ}} (T\acute{a}1 \not\equiv (S\acute{a} \,.\, \dot{S}))))$$

---

*The choice of the second expression $(Ta1 \mathrel{\text{þ}} T\acute{a}1)$ above would lead to alternative definitions of the intermittent traces in which the after-traces have an inhibitory rather than excitatory influence on the action of the after-cue. These definitions would be appropriate to a hedonistic theory in which reward is dominant over punishment; namely, one in which the varying complex has an inhibitory instead of excitatory influence in the definition of the regular and intermittent traces.

Next occultation is defined in terms of variation and conjunction as in the second definition given in Chapter I.

$$Ta2 \equiv (T\acute{a}2 \neq (T\acute{a}1\ .\ (T\acute{a}1 \neq (T\acute{a}1 \neq (S\acute{a}\ .\ \dot{S})))))$$

The expression on the right of the theorem above is simplified as follows.

$$Ta2 \equiv (T\acute{a}2 \neq (T\acute{a}1\ .\ (S\acute{a}\ .\ \dot{S})))$$

Rearrange the variables on the right and omit inner parentheses.

$$Ta2 \equiv (T\acute{a}2 \neq (S\acute{a}\ .\ \dot{S}\ .\ T\acute{a}1)) \quad \text{QED}$$

The definition of Ta3 above in terms of variation and conjunction may be proved in the same way.

A similar definition of Tb2 follows.

$$Tb2 \equiv (T\acute{b}2 \neq (S\acute{b}\ .\ \dot{S}\ .\ T\acute{b}1))$$

Definitions of intermittent traces on the lines above though orderly become rather bulky and may not truly reflect mechanisms in the brain. More practical definitions in terms of variation and conjunction follow.

$$Ta2 \equiv (T\acute{a}2 \neq T\acute{a}1 \neq (T\acute{a}1\ .\ Ta1))$$

$$Ta3 \equiv (T\acute{a}3 \neq T\acute{a}2 \neq (T\acute{a}2\ .\ Ta2))$$

Another possibility arises with the second definition of occultation.

$$Ta2 \equiv (T\acute{a}2 \neq (T\acute{a}1\ .\ (T\acute{a}1 \neq Ta1)))$$

$$Ta3 \equiv (T\acute{a}3 \neq (T\acute{a}2\ .\ (T\acute{a}2 \neq Ta2)))$$

Finally the third definition of occultation in terms of variation and disjunction gives us the following formulae.

$$Ta2 \equiv (T\acute{a}2 \neq (T\acute{a}1 \vee Ta1) \neq Ta1)$$

$$Ta3 \equiv (T\acute{a}3 \neq (T\acute{a}2 \vee Ta2) \neq Ta2)$$

Trace-interaction

Now that the regular and intermittent traces have been defined we can write the trace-interaction which underlies the animal's repertoire of responses. The three traces Ta1, Ta2 and Ta3 are independent variables, so that the presence or absence of one does not entail the presence or absence of the others. It is possible to conceive of eight permutations of the two values. The definitions of comprehensive traces below follow a similar pattern to the definitions of cues.

Ta3	Ta2	Ta1							
0	0	0	Ta[0] ≡ (((((((S_1 v S_2 v S_3)	v Ta3)	≢ Ta3	v Ta2)	≢ Ta2)	v Ta1)	≢ Ta1)
0	0	1	Ta[1] ≡ ((((((S_1 v S_2 v S_3)	v Ta3)	≢ Ta3)	v Ta2)	≢ Ta2)		. Ta1)
0	1	0	Ta[2] ≡ ((((((S_1 v S_2 v S_3)	v Ta3)	≢ Ta3)		. Ta2)	v Ta1)	≢ Ta1)
0	1	1	Ta[3] ≡ (((((S_1 v S_2 v S_3)	v Ta3)	≢ Ta3)		. Ta2)		. Ta1)
1	0	0	Ta[4] ≡ ((((((S_1 v S_2 v S_3)		. Ta3)	v Ta2)	≢ Ta2)	v Ta1)	≢ Ta1)
1	0	1	Ta[5] ≡ (((((S_1 v S_2 v S_3)		. Ta3)	v Ta2)	≢ Ta2)		. Ta1)
1	1	0	Ta[6] ≡ (((((S_1 v S_2 v S_3)		. Ta3)		. Ta2)	v Ta1)	≢ Ta1)
1	1	1	Ta[7] ≡ ((((S_1 v S_2 v S_3)		. Ta3)		. Ta2)		. Ta1)

The comprehensive traces defined above are contrary, so that only one can have an excitatory influence on the action of the cue Sa at once. As a consequence, the stimuli of action are also contrary. Only one response can be made at once — the key to the organization of behaviour. Examples of theorems of contrariety follow.

$$Ta[0] \mid Ta[1]$$

$$Ta[1] \mid Ta[2]$$

$$Ta[0] \mid Ta[2]$$

The sequence of responses in the animal's repertoire, and even its size, may be different for different cues. The cue Sb, for example, may elicit a repertoire of four responses in the order [8], [5], [0] and [3]. The response [8] is one which did not figure in the repertoire of Sa.

Tb2	Tb1					
0	0	Tb[8] ≡ (((((S_1 v S_2 v S_3)	v Tb2)	$\not\equiv$ Tb2)	v Tb1)	$\not\equiv$ Tb1)
0	1	Tb[5] ≡ ((((S_1 v S_2 v S_3)	v Tb2)	$\not\equiv$ Tb2)		. Tb1)
1	0	Tb[0] ≡ ((((S_1 v S_2 v S_3)		. Tb2)	v Tb1)	$\not\equiv$ Tb1)
1	1	Tb[3] ≡ (((S_1 v S_2 v S_3)		. Tb2)		. Tb1)

Clearly Tb[8] is contrary to Tb[5], Tb[5] to Tb[0] and Tb[8] to Tb[0], etc. But comprehensive traces appropriate to different cues are not necessarily contrary. For instance, Ta[0] is not necessarily contrary to Tb[8], nor Ta[5] to Tb[5], nor Ta[0] to Tb[3], etc.

The receptacle in the two tables of trace-interaction above consists of a disjunction of the three elements, as in the case of definitions of cues in Chapter II. The receptacle is present except when no element and hence no cue is present. In that rare event the response will be one of instinctive inaction.

An Example of Learning

In the table overleaf the regular and intermittent traces are defined in terms of their after-effects, the after-cue Sá and the varying complex $\mathring{S}$. The definitions are given in full on page 41.

Sa	1	0	1	1	0	1	0	1	1	1	0	0	0	1	1	0	0	1	1	1
Sá	0	1	0	1	1	0	1	0	1	1	1	0	0	0	1	1	0	0	1	1
Ṡ(pain or need)	1	0	0	1	1	1	1	0	1	1	0	1	1	0	0	1	0	0	1	1
(Sá . Ṡ)	0	0	0	1	1	0	1	0	1	1	0	0	0	0	0	1	0	0	1	1
Ta1	0	0	0	1	0	0	1	1	0	1	1	1	1	1	1	0	0	0	1	0
Tá1	0	0	0	0	1	0	0	1	1	0	1	1	1	1	1	1	0	0	0	1
(Sá . Ṡ. Tá1)	0	0	0	0	1	0	0	0	1	0	0	0	0	0	0	1	0	0	0	1
Ta2	0	0	0	0	1	1	1	1	0	0	0	0	0	0	0	1	1	1	1	0
Tá2	0	0	0	0	0	1	1	1	1	0	0	0	0	0	0	0	1	1	1	1
(Sá . Ṡ. Tá1 . Tá2)	0	0	0	0	0	0	0	0	1	0	0	0	0	0	0	0	0	0	0	1
Ta3	0	0	0	0	0	0	0	0	1	1	1	1	1	1	1	1	1	1	1	0
Tá3	0	0	0	0	0	0	0	0	0	1	1	1	1	1	1	1	1	1	1	1
Ta[ ]	[0]	[0]	[0]	[1]	[2]	[2]	[3]	[3]	[4]	[5]	[5]	[5]	[5]	[5]	[5]	[6]	[6]	[6]	[7]	[0]
Summary																				
Example 9	1	2	3	4	5	6	7	8	9	10	11	12	13	14	15	16	17	18	19	20
Sa elicits	[0]	*	[0]	[1]	*	[2]	*	[3]	[4]	[5]	*	*	*	[5]	[5]	*	*	[6]	[7]	[0]
Ṡ(pain or need)	1	0	0	1	1	1	1	0	1	1	0	1	1	0	0	1	0	0	1	1

Only the values of Sa and $\dot{S}$ need be known to determine the values of all the other variables. It is assumed that the after-effects in the beginning are all absent.

The comprehensive traces Ta[ ] are defined in terms of the regular and intermittent traces, according to definitions on page 43. The interaction is summarized by a behaviour and experience series which obey the rules laid down in drawing up eight similar examples at the beginning of the chapter.

Chapter IV

FEELINGS

Need and its Satisfaction

A stimulus of need, such as hunger or thirst, is a member of the disjunction called the varying complex. It therefore entails the complex. If we represent a stimulus of need by the expression $(S_1 \not\supset S_2)$ and the varying complex by the usual symbol $\mathring{S}$, we can draw up a theorem of entailment in line with the argument at the end of Chapter I, thus.

$$(S_1 \not\supset S_2) \supset \mathring{S}$$

The stimulus of need $(S_1 \not\supset S_2)$ depends on two factors; the presence of one stimulus and the absence of another. The second stimulus, represented by S_2 in the example above, corresponds to the object which satisfies the need, such as food to satisfy hunger or water to quench thirst. The stimulus of need is only present in the absence of this object; indeed, the object is logically contrary to the stimulus of need as a whole. This was also made clear in Chapter I where it was shown that the stimuli S_2 and $(S_1 \not\supset S_2)$ are contrary. In other words, it is impossible to experience the feeling of need elicited by $(S_1 \not\supset S_2)$ and the feeling of its satisfaction elicited by S_2 simultaneously. But it is of course possible to experience neither.

Also in Chapter I the operation of occultation was defined in terms of the commutative, associative operations of variation and conjunction. The first of these definitions stated that $(S_1 \not\supset S_2)$ and $(S_1 \ddagger (S_1 \,.\, S_2))$ are equivalent. We may compare this definition with the analogous definition of the regular trace, Ta1.

$$Ta1 \equiv (T\acute{a}1 \ddagger (S\acute{a} \,.\, \mathring{S})) \qquad \text{Primary variation}$$

$$(S_1 \not\supset S_2) \equiv (S_1 \ddagger (S_1 \,.\, S_2)) \qquad \text{Secondary variation}$$

The similarity of the theorems above is striking. The signs of variation and conjunction in the expressions on the right are the same, and the position of the parentheses is identical. In the definition of the regular trace we spoke of primary variation, while, by the analogy above, we speak of secondary variation in the definition of the stimulus of need. We also speak of primary stimuli of reaction of need eliciting primary feelings of need, and of secondary stimuli of reaction of satisfaction eliciting secondary feelings of satisfaction.

A stimulus of need has a punishing effect on behaviour. Its presence entails the presence of the varying complex and this has an excitatory influence on the action of the after-cue to bring about primary variation. Under these conditions the animal makes a variety of responses to the same cue. But the object which satisfies the need has an inhibitory influence on the effect of the stimulus of need and so acts as a reward. The response is then repeated. Learning of this kind is called appetitive.

Pain

A pain stimulus is a primary stimulus of reaction which elicits a primary feeling. In this respect it is similar to a stimulus of need. But unlike need a pain stimulus has no contrary.

The element S_4, for example, may be a pain stimulus and entail the varying complex $\hat{S}$, as in the theorem which follows.

$$S_4 \supset \hat{S}$$

There are two kinds of learning connected with pain. In one the animal endeavours to escape from a painful situation while in the other it attempts to avoid it altogether. In escape learning, then, the pain stimulus would be present as a member of the positive association of the cue before a response, while in avoidance learning it would be an absent member of the negative association. But in both kinds of learning this stimulus has a punishing effect when it follows an action, for the animal subsequently makes a different response to the cue. Once again, the alteration in behaviour depends on primary variation.

Fear of Pain

Avoidance conditioning is a process by which neutral elements acquire the punishing power of a pain stimulus. Suppose, for instance, a cue elicits a response which is followed by pain. Suppose that the animal makes other responses to this cue but each one in turn is followed by the pain stimulus. It is to the advantage of the animal if some element in the cue acquires punishing power, so that the situation it defines is itself avoided.

In the example below the unconditioned pain stimulus is S_4, while the elements S_5 and S_6 only become conditioned stimuli when the traces T_5 and T_6 are present. They then become conditioned primary stimuli of reaction of fear of pain eliciting primary feelings of fear of pain and so entail the presence of the varying complex.

$$\{ \quad S_4 \vee (S_5 \,.\, T_5) \vee (S_6 \,.\, T_6) \ \} \quad \supset \$$$

The element S_5 in the theorem of entailment above acquires punishing power when T_5 is present. The trace T_5 is defined in terms of its after-effect T_5', the after-effect S_5' and the pain stimulus S_4, as follows.

$$T_5 \equiv ((T_5' \not\equiv (S_5' \,.\, S_4)) \not\equiv (S_5' \,.\, T_5'))$$

To demonstrate the truth of the definition of T_5 above let us assume first that the value of S_5' is presence, represented by the digit 1.

If $S_5' \equiv 1$

then $T_5 \equiv ((T_5' \not\equiv (1 \,.\, S_4)) \not\equiv (1 \,.\, T_5'))$

and $T_5 \equiv ((T_5' \not\equiv S_4) \not\equiv T_5')$

and $T_5 \equiv S_4$

The trace T_5 takes the value of the unconditioned pain stimulus S_4. Let us now assume the alternative, that the value of S_5' is absence, represented by the digit 0.

If $S_5' \equiv 0$

then $T_5 \equiv ((T_5' \not\equiv (0 \,.\, S_4)) \not\equiv (0 \,.\, T_5'))$

and $T_5 \equiv ((T_5' \not\equiv 0) \not\equiv 0)$

and $T_5 \equiv T_5'$

The trace T_5 thus persists between successive presentations of S_5.

Second-order conditioning occurs when the neutral element acquires the punishing power of another neutral element which itself has become a conditioned stimulus. Avoidance of a painful situation even further ahead in time is then made possible. In the theorem of entailment above the neutral element S_6 becomes a conditioned stimulus when the trace T_6 is present. This trace is defined as follows.

$$T_6 \equiv ((T'_6 \not\equiv (S'_6 \,.\, (S_4 \vee (S_5 \,.\, T_5)))) \not\equiv (S'_6 \,.\, T'_6))$$

Higher-order conditioning on the same lines ensures the breaking-up of chains of responses leading to a painful conclusion.

As in the case of trial-and-error learning this account of conditioning goes beyond the simple teleological explanation which is usually offered. Tables of successive values could be constructed which show how the mechanisms underlying the process work. An example is given below. S_5 becomes a conditioned stimulus in space 6. Experimental extinction takes place in space 12.

	1	2	3	4	5	6	7	8	9	10	11	12	13	14	15
S_5	0	0	1	0	1	0	0	1	0	0	1	0	1	1	0
S'_5	0	0	0	1	0	1	0	0	1	0	0	1	0	1	1
S_4	0	0	0	0	0	1	0	0	1	0	0	0	0	0	0
$(S'_5 \,.\, 4)$	0	0	0	0	0	1	0	0	1	0	0	0	0	0	0
T_5	0	0	0	0	0	1	1	1	1	1	1	0	0	0	0
T'_5	0	0	0	0	0	0	1	1	1	1	1	1	0	0	0
$(T'_5 \not\equiv (S'_5 \,.\, S_4))$	0	0	0	0	0	1	1	1	0	1	1	1	0	0	0
$(S'_5 \,.\, T'_5)$	0	0	0	0	0	0	0	0	1	0	0	1	0	0	0
Summary	1	2	3	4	5	6	7	8	9	10	11	12	13	14	15
S_5	0	0	1	0	1	0	0	1	0	0	1	0	1	1	0
S_4	0	0	0	0	0	1	0	0	1	0	0	0	0	0	0
T_5	0	0	0	0	0	1	1	1	1	1	1	0	0	0	0

Hope for Satisfaction of a Need

Appetitive conditioning is a process by which a neutral element acquires the rewarding power of an object which satisfies a need. Suppose, for instance, a cue elicits a response which is followed by the satisfaction of a need. Suppose that the animal makes other responses to this cue and each one in turn is followed by the object of satisfaction. It is to the advantage of the animal if some element in the cue acquires rewarding power, so that the situation it defines is itself sought after.

In the theorem of entailment below the unconditioned object of satisfaction is S_8, while the neutral element S_9 becomes a conditioned stimulus if the trace T_9 is present. It then becomes a conditioned secondary stimulus of reaction of hope for satisfaction eliciting a secondary feeling of hope for satisfaction.

$$(S_7 \not\triangleright \{ S_8 \vee (S_9 \, . \, T_9) \}) \supset \mathring{S}$$

The element S_9 in the theorem above acquires rewarding power when T_9 is present. The trace T_9 is defined in exactly the same way that the traces in avoidance conditioning were defined, in terms of its after-effect T_9', the after-effect S_9' and the object of satisfaction S_8, as follows.

$$T_9 \equiv ((T_9' \not\equiv (S_9' \, . \, S_8)) \not\equiv (S_9' \, . \, T_9))$$

When the element S_9 acquires the rewarding power of the object it becomes contrary to the stimulus of need. An animal cannot feel both a need and hope for its satisfaction at one and the same time.

Higher-order conditioning ensures the formation of chains of responses leading to the satisfaction of a need. The mechanisms underlying the process are identical to those which serve in the case of higher-order avoidance conditioning.

Occultation in the theorem of entailment above can be defined in terms of the commutative, associative operations of variation and conjunction. The first definition in Chapter I is appropriate again here, and even more so when large numbers of neutral elements are involved.

The field of the three elements S_1, S_2 and S_3, in which only seven cues may be defined, is severely restricted and one cannot give examples of many kinds of learning. A field of nine elements, including S_1, S_4 and S_7 as primary, S_2 and S_8 as secondary and S_3, S_5, S_6 and S_9 as neutral, would give rise to five hundred and eleven cues.* The primary elements are entailed by primary stimuli of reaction and the secondary elements by secondary stimuli of reaction. It would be possible to show discrimination between the two needs and escape and avoidance learning taking place.

Emotions

So far we have tended to think about physical egocentric sensations of dissatisfaction (need), satisfaction and pain. But the symbols on previous pages could equally stand for emotions. Primary emotional dissatisfactions like anger, shame and sorrow are contrary to secondary emotional satisfactions like pleasure, pride and joy. Pleasure is mentioned here to establish its secondary status in a theory of learning which is not hedonistic. Hurt feelings on the other hand are the primary emotional equivalent of physical pain.

The dichotomy between mind and body is quite distinct from that between the psychological and material. Otherwise a concept like physical feeling would be a contradiction in terms. Emotions are elicited for the most part by visual and auditory stimuli of reaction which are material events. The mechanics of goal-seeking or avoidance behaviour apply therefore equally to them as to sensations. The elements which form emotional stimuli are simply far more complex logically than those behind physical feelings. Their existence must always be suspected when learning takes place in the absence of any discernible physical motivation.

---

*This number is reduced if the elements are not independent. Suppose two of them are made up as follows, for instance.

$$S_1 \equiv S_I$$
$$S_7 \equiv (S_{II} \nsupset S_I)$$

S_1 and S_7 are then contrary.

$$S_1 \mid S_7$$

The one hundred and twenty-eight cues in which both S_1 and S_7 are present would be logically impossible.

Dominance

The varying complex is a disjunction of stimuli of pain, need and fear of pain, any one of which is sufficient to bring about variation in response. In other words, the complex is entailed by one or more primary stimuli of reaction.

It is quite possible for a stimulus of pain or fear of pain to be present with an object of need-satisfaction or stimulus of hope for need-satisfaction. Equally, it is possible for a stimulus of one need to be present with an object or stimulus of hope for an object relevant to another. When either of these exigencies occur the presence of the primary stimulus entails the presence of the complex, regardless of the presence of the irrelevant secondary stimuli. For this reason feelings of pain, need and fear of pain are said to be dominant over feelings of satisfaction or hope for satisfaction. The secondary feelings are only incidental to the learning process.

The dominance of punishment over reward is brought about by the evolution of a logical nervous system in which the varying complex has an excitatory influence on the action of the after-cue in the definition of regular and intermittent traces. Under this system a response which results in the absence of all feeling is repeated when the same situation arises again.*

The varying complex is defined below. It contains the two stimuli of need, the pain stimulus and the two stimuli of fear. The element S_3 is not included in this particular example.

$$\mathring{S} \equiv ((S_1 \not\supset S_2) \vee S_4 \vee (S_5 \,.\, T_5) \vee (S_6 \,.\, T_6) \vee (S_7 \not\supset (S_8 \vee (S_9 \,.\, T_9))))$$

The inclusion of the elements S_1 and S_2 means that $(S_1 \not\supset S_2)$ and therefore the complex $\mathring{S}$ is entailed by the cues Sc and Sg, as in the theorems which follow.

$$Sc \supset \mathring{S}$$

$$Sg \supset \mathring{S}$$

---

*By way of a complete contrast an alternative, hedonistic system may be envisaged in which the varying complex has an inhibitory influence on the action of the after-cue. Rewards would then be dominant, punishments incidental and responses to the same cue would vary in the absence of all consequent feeling.

Chapter V

OTHER KINDS OF BEHAVIOUR

General Stimuli

Sometimes an animal responds to two or more cues in the same way. We then arrive at the notion of a general stimulus. The cues Sc and Sf, for example, may form the general stimulus (Sc v Sf). Either Sc or Sf entails (Sc v Sf).

Since Sc and Sf like other cues are contrary the sign of disjunction 'v' in the expression for the general stimulus may be replaced by the sign of variation '⫳', giving the form (Sc ⫳ Sf). The sign of disjunction means 'and/or', but when the possibility that both stimuli are present together is ruled out, as in the case of cues that are contrary, the sign of variation meaning 'or' in its exclusive sense may be used instead. Either mechanism is equally good.

In learning to respond to the general stimulus (Sc v Sf), punishment or reward of a response to the cue Sc has the same effect as punishment or reward of a response to Sf. To bring this about we need the concept of a common or shared comprehensive trace, which may be represented by the symbol Tcf[9] or Tcf[4], for example. For the sake of simplicity we shall assume that the repertoire of responses to (Sc v Sf) consists of only two actions, [9] and [4]. The theorems below show how the stimuli of action S[9] and S[4] are entailed by either cue and the common comprehensive traces.

$$((Sc \text{ v } Sf) \,.\, Tcf[9]) \supset S[9]$$

$$((Sc \text{ v } Sf) \,.\, Tcf[4] \supset S[4]$$

In order to define the comprehensive traces Tcf[9] and Tcf[4] we require first a definition of a common regular trace, which will be represented by the symbol Tcf1. As the repertoire consists of only two responses common intermittent traces are not required in this instance, although of course other examples could be given in which such intermittent traces were involved. The definition of the common regular trace follows.

$$Tcf1 \equiv (T\acute{c}f1 \not\equiv ((Sc \text{ v } Sf)' \,.\, \dot{S}))$$

An alternative definition of the common regular trace could be given in which the sign of disjunction is replaced by the sign of variation. Equally, the after-effect of the general stimulus $(Sc \text{ v } Sf)'$ or $(Sc \not\equiv Sf)'$ could be represented by the equivalent forms $(S\acute{c} \text{ v } S\acute{f})$ or $(S\acute{c} \not\equiv S\acute{f})$.

We are now able with the symbols above to define the common comprehensive traces Tcf[9] and Tcf[4].

Tcf1	
0	$Tcf[9] \equiv (((S_1 \text{ v } S_2 \text{ v } S_3) \text{ v } Tcf1) \not\equiv Tcf1)$
1	$Tcf[4] \equiv ((S_1 \text{ v } \dot{S}_2 \text{ v } S_3) \qquad .\ Tcf1)$

The receptacle $(S_1 \text{ v } S_2 \text{ v } S_3)$ is the usual one, a disjunction of the three elements.

General stimuli formed by the disjunction or variation of more than two cues are also conceivable.

Diversity of Responses with the Same Aim

Rather more difficult than the problem of simple generalization is that of explaining how an animal makes *different* responses to different cues which are directed towards the same end. If, for example, a certain response to a cue meets with punishment or reward as the case may be, by what mechanisms does the animal take heed that another response to a different cue is likely to meet with the same outcome? The problem is similar to that of simple generalization in that common comprehensive traces are involved.

Such traces are represented in this instance by the symbols Td[10]e[14], Td[11]e[15], etc. The presence of the common comprehensive trace Td[10]e[14], for example, indicates that the animal has made a decision to respond to the cue Sd by the action [10] *or* to respond to the cue Se by the action [14]. If either of these actions goes unpunished the common comprehensive trace Td[10]e[14] remains present and the decision accordingly is unchanged. On the other hand, if either action does meet with punishment the common comprehensive trace Td[10]e[14] becomes absent and the next common comprehensive trace in the series, which is Td[11]e[15], becomes present. A new decision to respond to the cue Sd by the action [11] or to respond to the cue Se by the action [15] has then been made. This decision will be implemented when either Sd or Se is next presented.

The theorems which follow demonstrate how the appropriate response to the cue is made. It will be seen that the same comprehensive trace figures in the second and third theorems, but different actions ensue, depending on which cue is present.

$$(\mathrm{Sd} \;.\; \mathrm{Td[10]e[14]}) \quad \supset \quad \mathrm{S[10]}$$

$$(\mathrm{Sd} \;.\; \mathrm{Td[11]e[15]} \quad \supset \quad \mathrm{S[11]}$$

$$(\mathrm{Se} \;.\; \mathrm{Td[11]e[15]}) \quad \supset \quad \mathrm{S[15]}$$

In this example the common regular trace Tde1 is used together with a single common intermittent trace Tde2. These are defined as follows.

$$\mathrm{Tde1} \equiv (\mathrm{T\acute{d}e1} \not\equiv ((\mathrm{Sd} \vee \mathrm{Se})' \;.\; \dot{S}))$$

$$\mathrm{Tde2} \equiv (\mathrm{T\acute{d}e2} \not\equiv ((\mathrm{Sd} \vee \mathrm{Se})' \;.\; \dot{S} \;.\; \mathrm{T\acute{d}e1}))$$

The interaction of the common regular and intermittent traces defined above to form the common comprehensive traces follows the usual pattern, as below.

Tde2	Tde1	
0	0	$\mathrm{Td[10]e[14]} \equiv (((((S_1 \vee S_2 \vee S_3) \vee \mathrm{Tde2}) \not\equiv \mathrm{Tde2}) \vee \mathrm{Tde1}) \not\equiv \mathrm{Tde1})$
0	1	$\mathrm{Td[11]e[15]} \equiv ((((S_1 \vee S_2 \vee S_3) \vee \mathrm{Tde2}) \not\equiv \mathrm{Tde2}) \;.\; \mathrm{Tde1})$
1	0	$\mathrm{Td[12]e[16]} \equiv ((((S_1 \vee S_2 \vee S_3) \;.\; \mathrm{Tde2}) \vee \mathrm{Tde1}) \not\equiv \mathrm{Tde1})$
1	1	$\mathrm{Td[13]e[17} \equiv (((S_1 \vee S_2 \vee S_3) \;.\; \mathrm{Tde2}) \;.\; \mathrm{Tde1})$

The responses [10], [11], [12] and [13] to the cue Sd are thus linked respectively with the responses [14], [15], [16] and [17] to the cue Se. Each pair of responses probably converge. The outcome of a response to one cue is likely to be the same as the outcome of the other response to the other cue.

Responses to more than two cues may also converge. The same end may be reached by responses to a whole set of cues. There is no difficulty symbolizing the mechanisms underlying this behaviour, extending the notation adopted above.

The intelligence of an animal is measured by its ability to generalize. We use the term to cover this kind of learning because the after-effect of a general stimulus (or after-effects of the cues in a general stimulus, which amounts to the same thing) figures in the definitions of the regular and intermittent traces. We call it complex generalization to distinguish it from the simple generalization discussed in the previous section.

Instinctive Actions

We have given examples of repertoires which consist of eight, four and two responses. If N is the number of regular and intermittent traces involved in the definitions of comprehensive traces, the number of responses in the repertoire is 2^N. The extreme case when the repertoire consists of only one response means that no regular or intermittent traces are involved. A single response of this kind may be regarded as instinctive. The theorem below shows that the cue Sg entails the stimulus of action S[5]. The action [5] in this example is an instinctive one and there is accordingly no symbol for a comprehensive trace.

$$Sg \equiv S[5]$$

We have also seen that the action [5] may be a response to Sa or Sb which is affected by experience. When however this action is elicited by the cue Sg punishment after the response has no effect on future behaviour. The extreme case of instinctive action is therefore one where a response which is punished is repeated when the same cue is presented again.

Definition of the Stimulus of Action

We are now able to define the stimuli of action in terms of effective cues, using information obtained from the various tables of comprehensive traces in preceding pages. As both cues and effective cues are contrary the sign of disjunction 'v' in the list below may be replaced throughout by the sign of variation '$\ddagger$'. Either mechanism will do.

$S[0] \equiv ((Sa \,.\, Ta[0]) \vee (Sb \,.\, Tb[0]))$

$S[1] \equiv (Sa \,.\, Ta[1])$

$S[2] \equiv (Sa \,.\, Ta[2])$

$S[3] \equiv ((Sa \,.\, Ta[3]) \vee (Sb \,.\, Tb[3]))$

$S[4] \equiv ((Sa \,.\, Ta[4]) \vee ((Sc \vee Sf) \,.\, Tcf[4]))$

$S[5] \equiv ((Sa \,.\, Ta[5]) \vee (Sb \,.\, Tb[5]) \vee Sg)$

$S[6] \equiv (Sa \,.\, Ta[6])$

$S[7] \equiv (Sa \,.\, Ta[7])$

$S[8] \equiv (Sb \,.\, Tb[8])$

$S[9] \equiv ((Sc \vee Sf) \,.\, Tcf[9])$

$S[10] \equiv (Sd \,.\, Td[10]e[14])$

$S[11] \equiv (Sd \,.\, Td[11]e[15])$

$S[12] \equiv (Sd \,.\, Td[12]e[16])$

$S[13] \equiv (Sd \,.\, Td[13]e[17])$

$S[14] \equiv (Se \,.\, Td[10]e[14])$

$S[15] \equiv (Se \,.\, Td[11]e[15])$

$S[16] \equiv (Se \,.\, Td[12]e[16])$

$S[17] \equiv (Se \,.\, Td[13]e[17])$

The contrariety of cues and effective cues implies that of the stimuli of action. Examples follow.

S[0] | S[1]

S[1] | S[2]

S[0] | S[2]

Serial Representation of Stimuli

When a set of stimuli are completely contrary, in the sense that only one can be present at once, a single series tells us which of them is present or absent at any given time. Cues, effective cues and stimuli of action may all be represented in this way. Cues, for instance, may be represented as usual by the seven symbols Sa to Sg as in the series below, with the digit 0 to denote the situation when all three elements are absent.

1	2	3	4	5	6	7	8	9	10	11	12	13	14	15	16
Sg	Sc	Sf	Sb	0	Sb	Sb	Se	Sd	Sg	Sa	Se	Se	Se	0	Sc

From this series we can tell, for example, that Sg is present in space 1 and absent in space 2, and so on. We can also see that Sb is present for the two consecutive spaces 6 and 7 and Se is present for the three consecutive spaces 12, 13 and 14. Otherwise cues are present for one space at a time.

Beneath this series let us now write one of stimuli of action. To save space the symbols S[0] to S[17] re shortened to [0] to [17], with the digit 0 to denote instinctive inaction. Under this is written an experience series, bearing in mind that the cues Sc and Sg both entail the varying complex $\dot{S}$.

	1	2	3	4	5	6	7	8	9	10	11	12	13	14	15	16
	Sg	Sc	Sf	Sb	0	Sb	Sb	Se	Sd	Sg	Sa	Se	Se	Se	0	Sc
	[5]	[9]	[4]	[8]	0	[5]	[0]	[14]	[11]	[5]	[0]	[16]	[17]	[14]	0	[9]
$\dot{S}$	1	1	1	1	1	1	1	1	1	1	0	0	1	1	0	1

The continued presence of the varying complex ensures a wide variety of stimuli of action in the middle series, which is entirely determined by the upper and lower ones. There are examples here of instinctive responses and both kinds of generalization, as well as simple trial-and-error learning.

The latter is displayed by successive responses to the cue Sb. Its first presentation in space 4 elicits the response [8]. As this action has an aversive outcome in space 5 a new response [5] is elicited in space 6. This also has an aversive outcome in space 7 and the next response in the repertoire, which is [0], is elicited in the same space, the effect of $\breve{S}$ in this case being immediate.

The cases of instinctive action and inaction, represented by the symbols [5] and 0 respectively, show that punishment does not necessarily produce a different response. For example, the cue Sg elicits the response [5] in space 1 which is repeated in space 10 despite punishment in space 2. Similarly, inaction in space 5 is repeated in space 15 despite punishment in space 6. But the response [5] may also be elicited by the cue Sb, as in space 6, when punishment does have the effect of bringing about a change.

The cue Sc elicits the response [9] in space 2, which is punished. Its next presentation in space 16 elicits the same response. We should however be wrong to conclude that the action is instinctive, since the second response in the repertoire is elicited by Sf in space 3. This action is [4] and also has an aversive outcome, so the cycle is completed. When the repertoire is small, as in this case, generalization often means that responses to cues may appear to be instinctive.

The responses to Sd and Se converge. Here we are dealing with two linked repertoires. Successively punished responses to each cue do not necessarily go in numerical order. For example, the response [14] to Se in space 8 is followed by the response [16] when the cue is next presented in space 12, since the response [11] to Sd intervenes. The response [15] is missed out. Simple trial-and-error learning follows in spaces 13 and 14.

In the serial representation above, and also in the behaviour and experience series drawn up in Chapter III, it has been assumed that each stimulus is present or absent for a whole number of spaces at a time. But this over-simplifies the issue. There is no reason why a stimulus should not be present or absent for a fraction of a space. In the series below, for instance, each space is divided into quarters.

	1				2				3				4			
	Sa	Sg	Sa	Sd	Sd	0	Sa	Sc	Sf	Sf	Sf	Sf	Sf	Sf	Sf	Sd
	[0]	[5]	[0]	[10]	[10]	0	[1]	[9]	[9]	[9]	[9]	[9]	[9]	[9]	[9]	[11]
Ṡ	0	1	0	0	0	0	1	1	0	0	1	1	0	0	0	0

The cue Sd is present for two consecutive quarters in spaces 1 and 2, while the cue Sf is present for seven consecutive quarters in spaces 3 and 4. In other words, Sd is present for half a space and Sf for one and three-quarter spaces. The other cues are present for a quarter of a space each.

Once again, the stimuli of action in the middle series are determined entirely by the upper and lower ones. Since Sa elicited [0] in the third quarter of space 1 the presence of the varying complex Ṡ in the third quarter of space 2 means that Sa elicits [1] in the same quarter of that space, the effect of Ṡ in this case being immediate. Similarly, since Sd elicited [10] in the fourth quarter of space 1 the presence of Ṡ in the fourth quarter of space 2 means that Sd elicits [11] in the fourth quarter of space 4.

Series could also be written in intervals of a half or a third of a space, or indeed in intervals of any fraction. Drawing them up helps to clarify the concept of a stimulus, especially with regard to the question of its duration. It may be present or absent for any rational number of spaces.

Chapter VI

PATHWAYS IN THE BRAIN

Operations

One way in which we may interpret the formulae in earlier pages is by drawing diagrams of the nervous system. Neurons, for example, may be represented as follows.

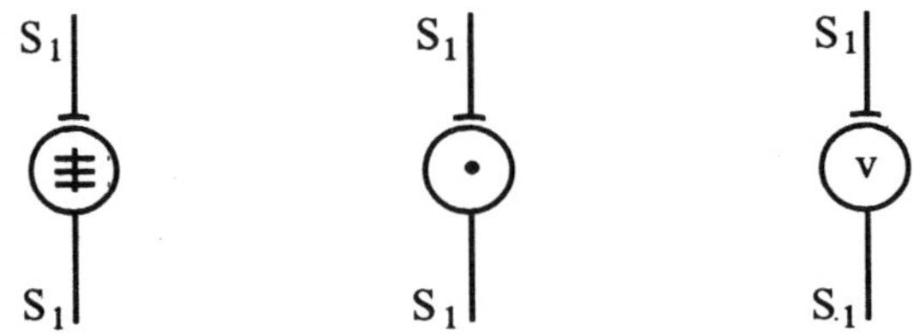

The lines represent nerve fibres and their synaptic connexions with the neurons are depicted thus:  . There is one-way transmission of impulses at the synapses which in the examples above is in a downwards direction.

The three fundamental commutative operations may be depicted as follows.

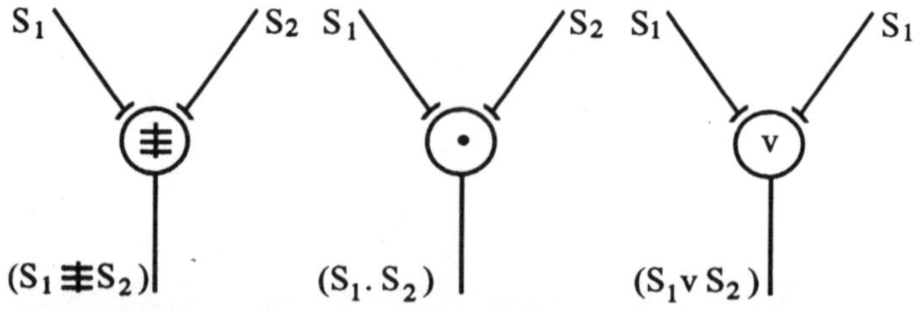

These operations are also associative. Associations of three stimuli are illustrated as follows.

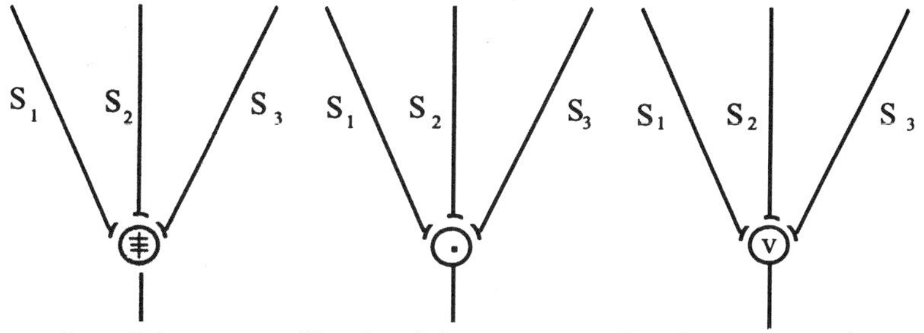

$(S_1 \not\equiv S_2 \not\equiv S_3)$ $(S_1 . S_2 . S_3)$ $(S_1 \vee S_2 \vee S_3)$

Associations of four or more elements may be represented in the same way. A variation is present when an odd number of elements is present, a conjunction is present when all its elements are present and a disjunction is present when at least one element is present.

The centre diagrams above show conjunction, which is the logical operation underlying the process of nervous excitation. Occultation, which is the logical operation underlying the process of nervous inhibition, is held to depend on two operations, since it is neither commutative nor associative. The three definitions of this operation in terms of commutative, associative operations are represented as follows. In each case two neurons are involved. There is bifurcation of fibres at S_1 in Def. (1) and Def. (2) and at S_2 in Def. (3). Once again, the diagrams indicate that impulses travel in a downwards direction.

Def (1) **Def (2)** **Def (3)**

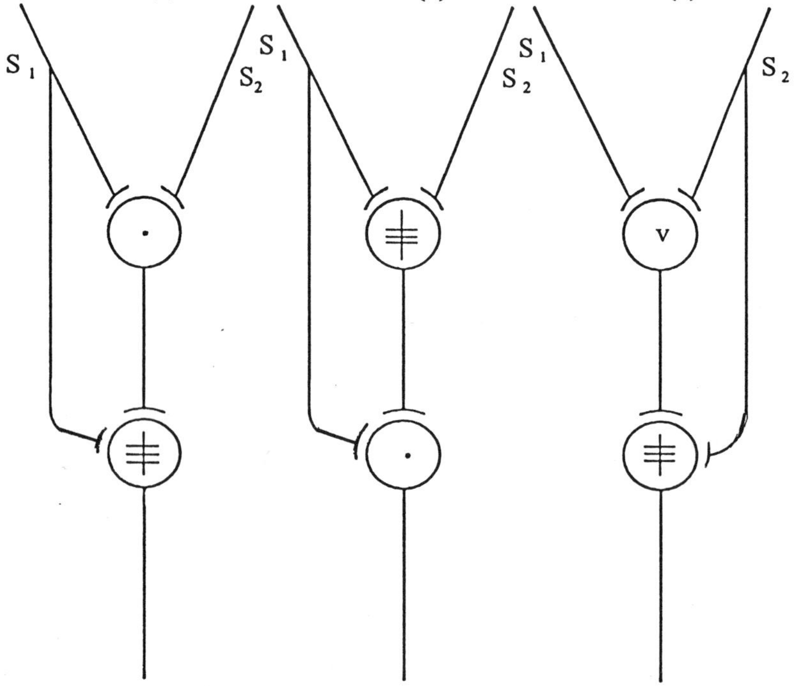

$(S_1 \not\equiv (S_1 \,.\, S_2))$ $(S_1 \,.\, (S_1 \not\equiv S_2))$ $((S_1 \text{ v } S_2) \not\equiv S_2)$

$(S_1 \not\supset S_2)$ $(S_1 \not\supset S_2)$ $(S_1 \not\supset S_2)$

The Formation of a Stimulus of Action

The diagram below maps the neurological structure underlying the formation of the cue Sa by interaction of the elements S_1, S_2 and S_3. As in electrical diagrams the convention of one fibre leaping over another when there is no connexion between them has been adopted, thus  . As far as possible this usage is kept to a minimum.

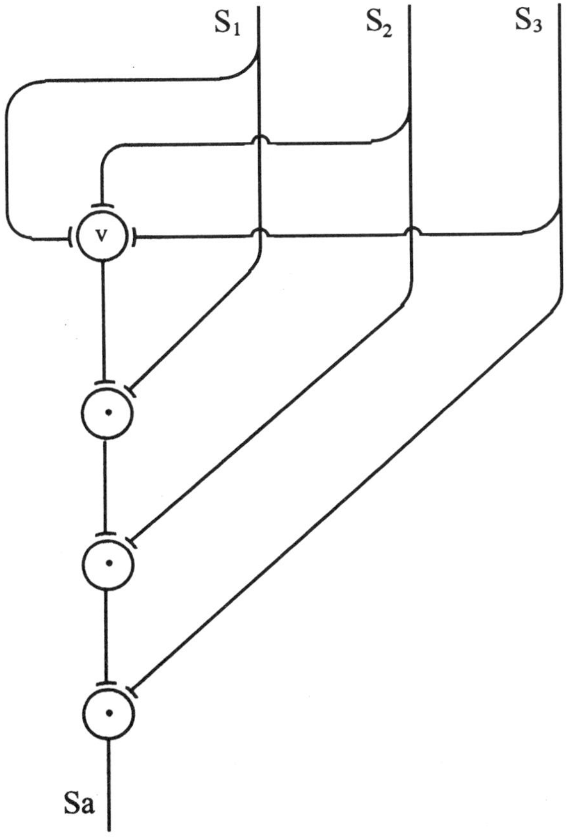

$$Sa \equiv ((((S_1 \text{ v } S_2 \text{ v } S_3) \,.\, S_1) \,.\, S_2) \,.\, S_3)$$

A basic function in the nervous system is the after-effect of a cue. It is represented in the following way. The arrow shows the direction of impulses, in this case downwards.

After-effects of regular and intermittent traces are represented in the same manner.

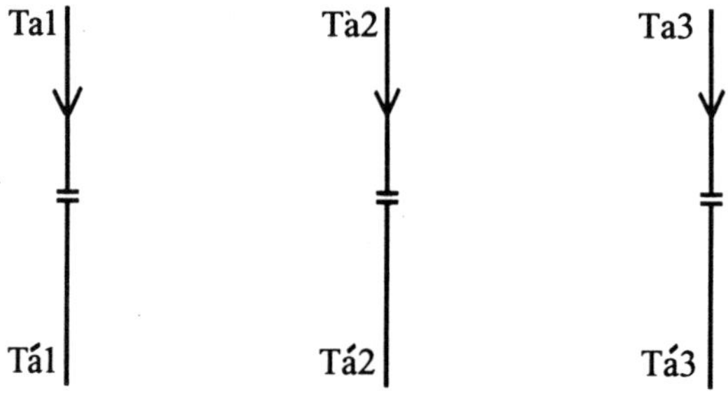

The after-cues and after-traces follow after a short space of time. Operations in comparison are more or less instantaneous. We may safely disregard the time taken in carrying them out.

The diagram below shows how the regular trace Ta1 and intermittent traces Ta2 and Ta3 are defined in terms of after-traces, the after-cue Sá and the varying complex Ŝ.

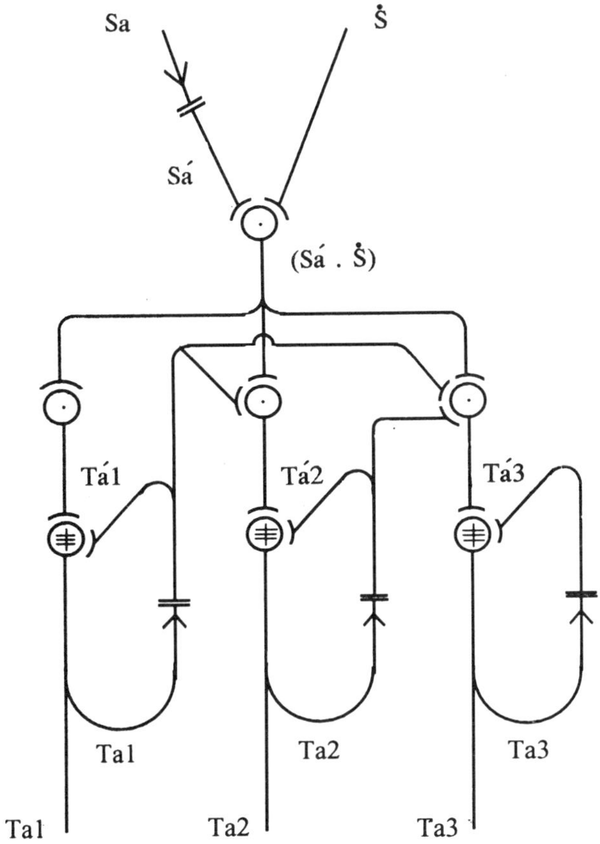

$$Ta1 \equiv (Ta'1 \not\equiv (Sa' \,.\, \dot{S}))$$

$$Ta2 \equiv (Ta'2 \not\equiv (Sa' \,.\, \dot{S} \,.\, Ta'1))$$

$$Ta3 \equiv (Ta'3 \not\equiv (Sa' \,.\, \dot{S} \,.\, Ta'1 \,.\, Ta'2))$$

The diagram above shows how circuits underlie the definitions of traces. The first circuit on the left is that of the regular trace while the second and third circuits are those of the intermittent traces. Impulses travel in an anti-clockwise direction in each case.

The next diagram shows how trace or motor interaction takes place. In this example we witness the formation of the comprehensive trace Ta[5]. Diagrams could be drawn up for any of the other seven comprehensive traces corresponding to this cue.

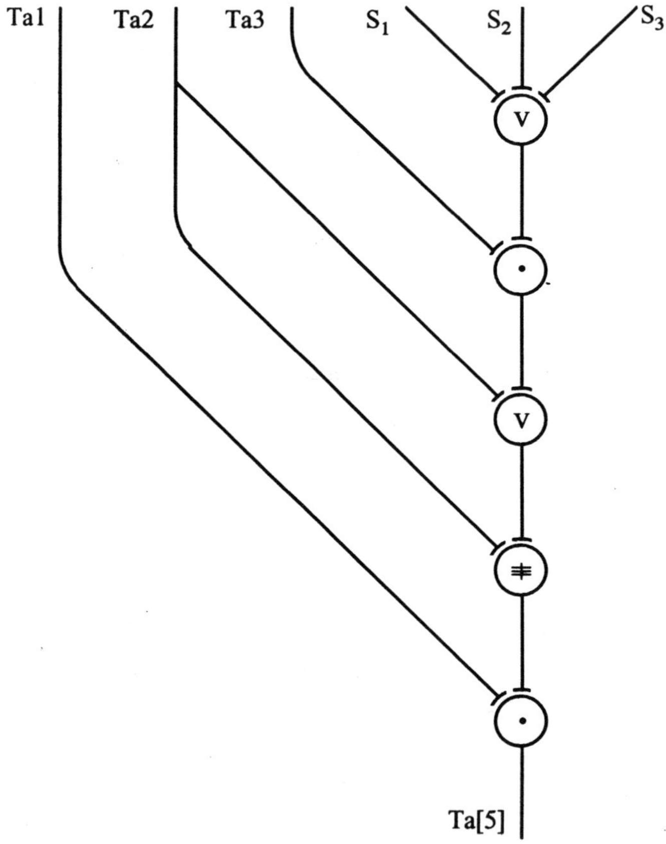

$$T[a]5 \equiv (((((S_1 \text{ v } S_2 \ S_3) \ . \ Ta3) \text{ v } Ta2) \ddagger Ta2) \ . \ Ta1)$$

The diagram below pictures the neurology underlying the formation of the cue Sb. Sensory interaction in this case is a little more complicated than in that of Sa.

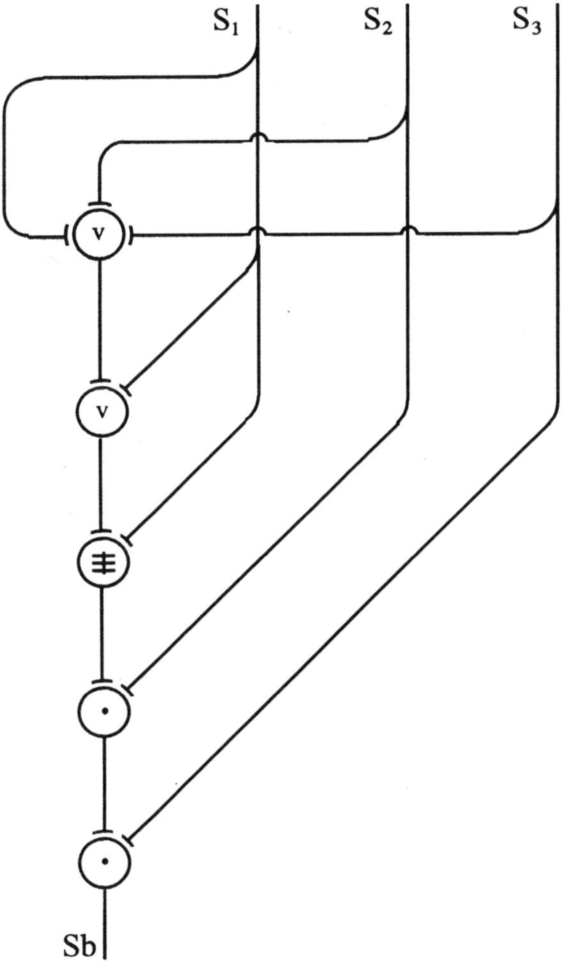

$Sb \equiv (((((S_1 \text{ v } S_2 \text{ v } S_3) \text{ v } S_1) \not\equiv S_1) \,.\, S_2) \,.\, S_3)$

The diagram which follows illustrates how the regular trace Tb1 and the intermittent trace Tb2 are defined. Once again, each trace is formed by means of a circuit, but in this case there are only two.

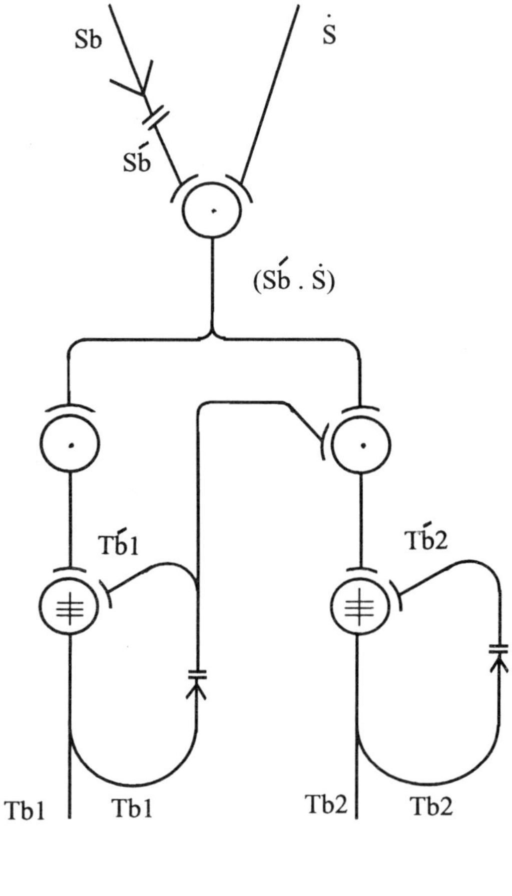

$$Tb1 \equiv (T\acute{b}1 \ddagger (S\acute{b} \,.\, \dot{S}))$$

$$Tb2 \equiv (T\acute{b}2 \ddagger (S\acute{b} \,.\, \dot{S} \,.\, T\acute{b}1))$$

The cue Sb has a repertoire of four responses. To continue our argument we show in the diagram below how the comprehensive trace Tb[5] is formed. Similar diagrams could be drawn which depict how the other three comprehensive traces are defined.

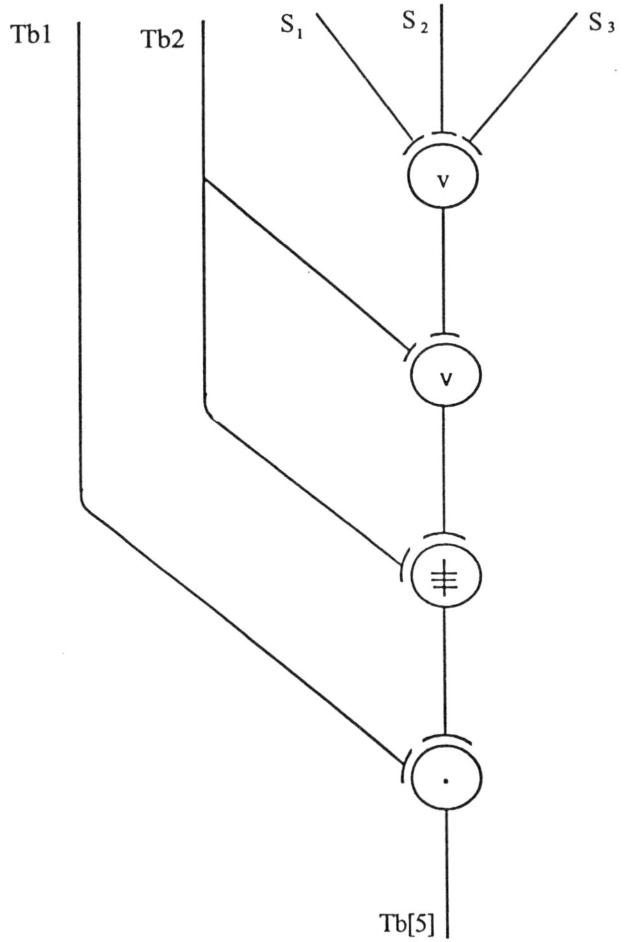

$$Tb[5] \equiv ((((S_1 \text{ v } S_2 \text{ v } S_3) \text{ v } Tb2) \not\equiv Tb2) \; . \; Tb1)$$

Finally, the neurological structure which underlies the definition of the cue Sg is represented in the diagram below.

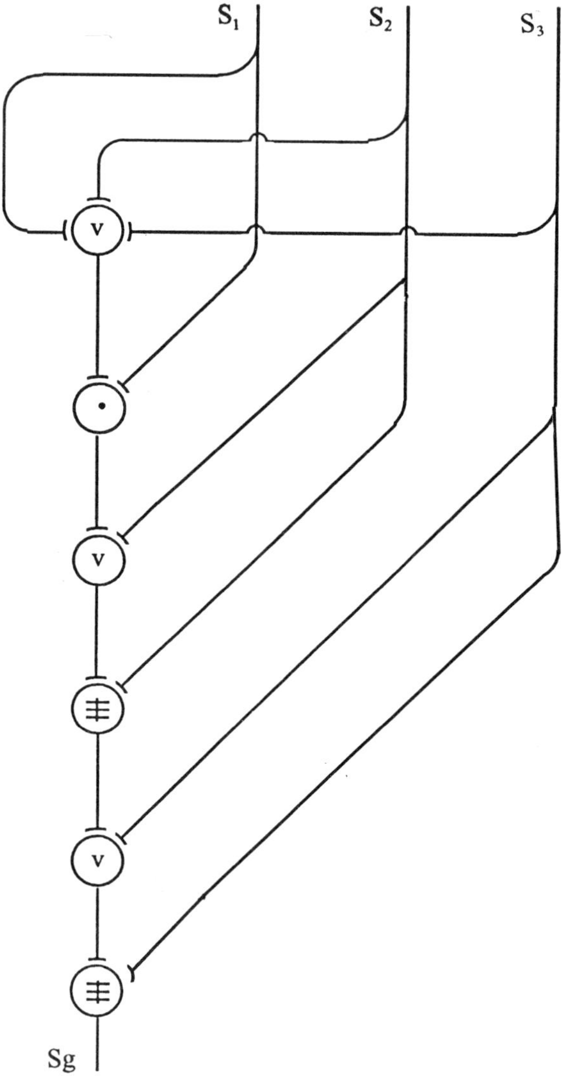

$$Sg \equiv ((((((S_1 \lor S_2 \lor S_3) \,.\, S_1) \lor S_2) \not\equiv S_2) \lor S_3) \not\equiv S_3)$$

We are now able to complete the definition of the stimulus of action S[5] in terms of the cues Sa and Sb and their comprehensive traces Ta[5] and Tb[5]. The cue Sg also figures in the definition.

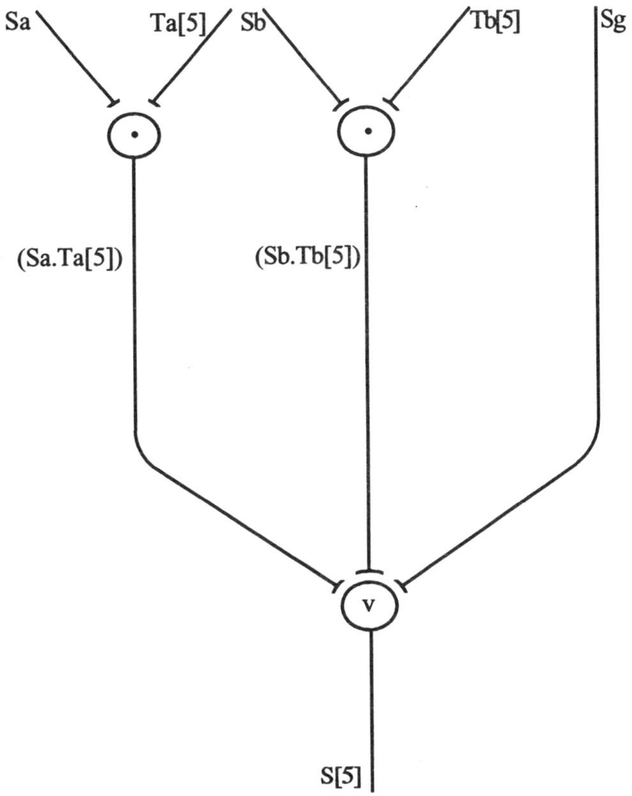

S[5] ≡ ((Sa . Ta[5]) v (Sb . Tb[5]) v Sg)

Generalization

The problem of generalization in its most simple form is that of explaining how an animal responds in the same way to two or more cues. There is no intrinsic reason why these should be similar, although it may often happen in practice. To stress this point the example we gave in the previous chapter used cues which were as different as possible, the cue Sc being the obverse of Sf.

The repertoire of the general stimulus (Sc v Sf) contains only two responses. The motor interaction on which they depend involves a regular but no intermittent traces. This is defined by a single circuit, as in the following diagram.

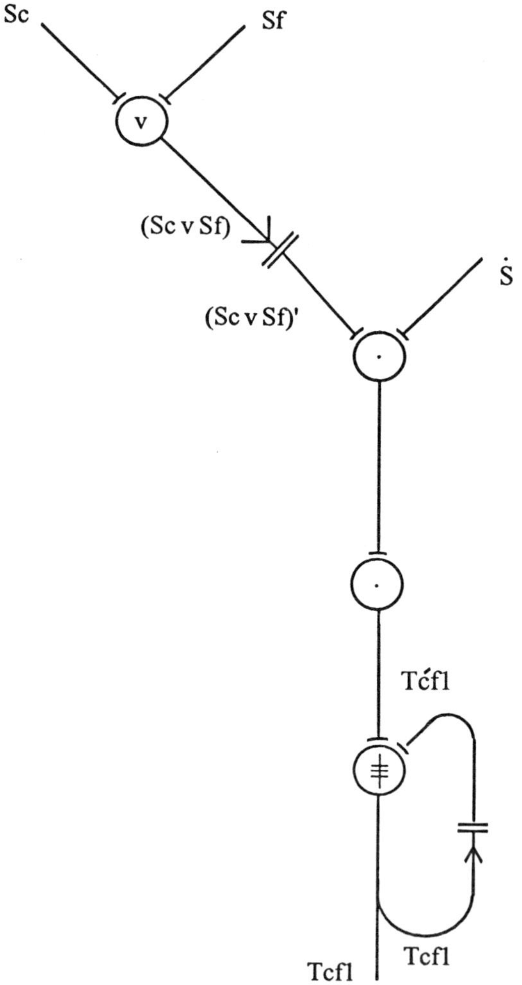

$$Tcf1 \equiv (T\acute{c}f1 \not\equiv ((Sc \text{ v } Sf)' \;.\; \dot{S}))$$

The absence or presence of the regular trace Tcf1 determines whether the next response to the general stimulus (Sc v Sf) will be [9] or [4] respectively.Let us consider the comprehensive trace Tcf[9]. The diagram which follows shows how the stimulus of action S[9] is defined.

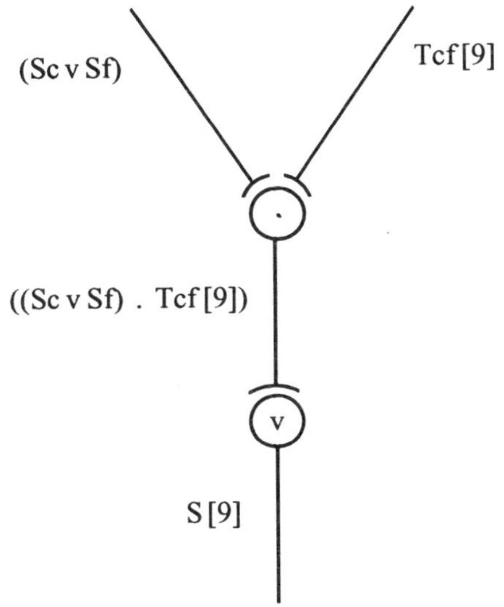

S [9] ≡ ((Sc v Sf) . Tcf [9])

Far more important than the simple form of generalization considered above is that of the cues Sd and Se. As in the case of Sc and Sf the cues were chosen because they are quite dissimilar, Sd being the obverse of Se. But unlike the example above responses to Sd differ from those to Se, though it is assumed that the same end is reached by the response [10] to Sd and the response [14] to Se, for instance. The same is true of the response [11] to Sd and the response [15] to Se, and so on. This phenomenon is convergence. On a grand scale it explains how an animal is able to respond intelligently to a host of situations never encountered before, on the basis of information derived from comparatively few experiences.

The mechanisms underlying convergence are pictured in the diagrams below. In this kind of generalization the after-cues are defined separately, although strictly speaking the same result would be achieved using the after-effect of their disjunction (Sd v Se)′.

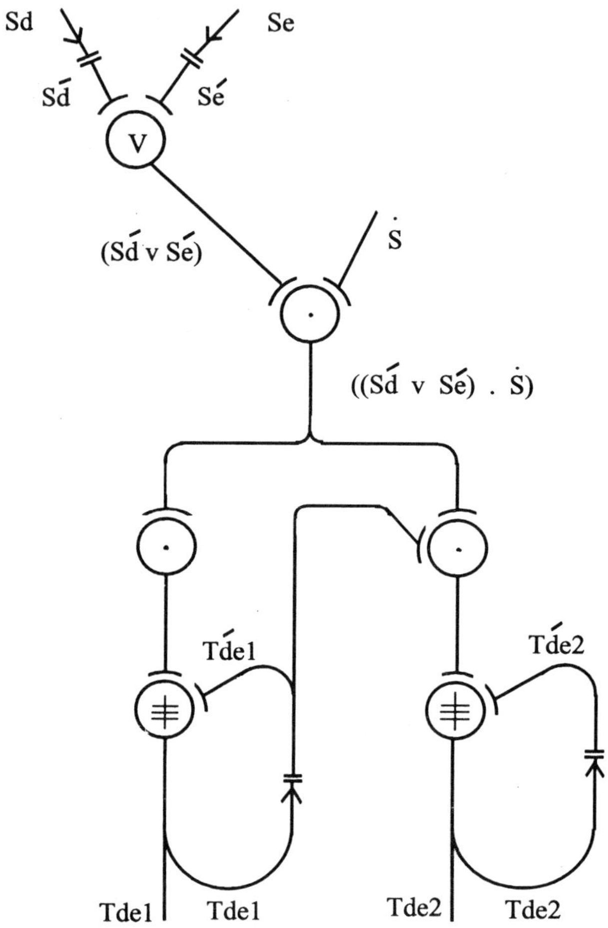

$$Tde1 \equiv (T\acute{d}e1 \not\equiv ((S\acute{d} \vee S\acute{e}) \,.\, \dot{S}))$$

$$Tde2 \equiv (T\acute{d}e2 \not\equiv ((S\acute{d} \vee S\acute{e}) \,.\, \dot{S} \,.\, T\acute{d}e1))$$

The diagrams below show how the comprehensive traces Td[10]e[14] and Td[11]e[15], which are defined in terms of the traces Tde1 and Tde2, form different stimuli of action depending on which cue is present.

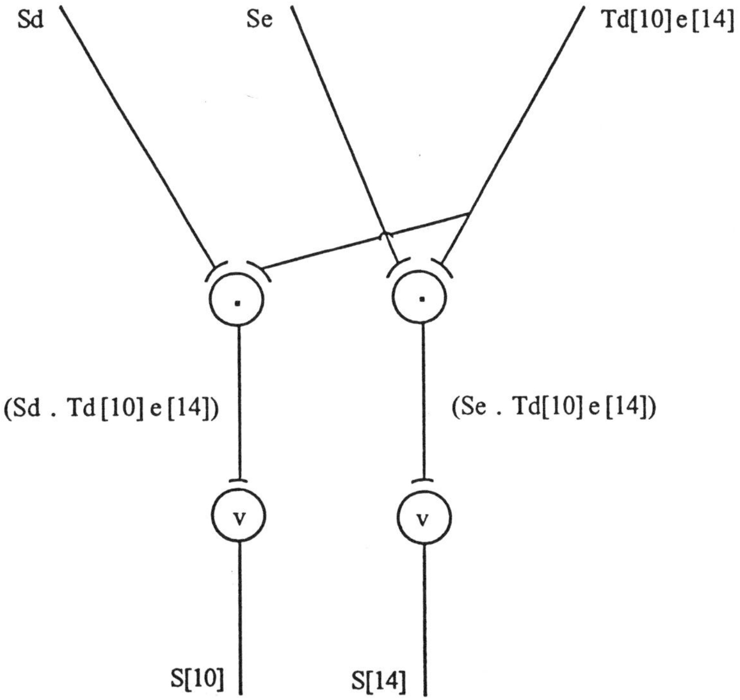

S[10] ≡ (Sd . Td[10]e[14])

(S[14] ≡ (Se . Td[10]e[14])

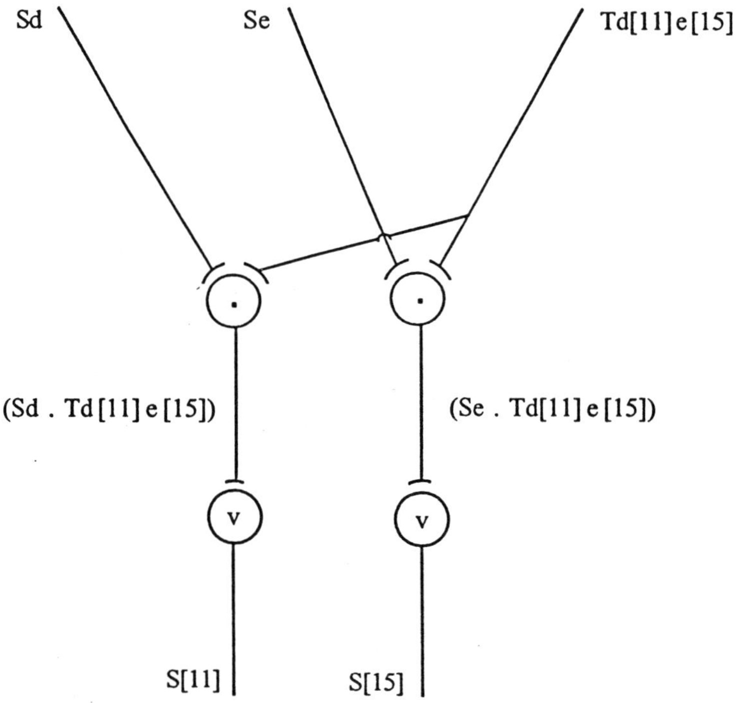

S[11] ≡ (Sd . Td[11]e[15])

S[15] ≡ (Se . Td[11]e[15])

Conditioning

Like trial-and-error learning conditioning depends on mechanisms arranged in the form of circuits. But the diagram of a conditioning trace is more complicated than that of a regular trace, as we should expect from a study of their respective formulae. It includes two integrated circuits as illustrated below.

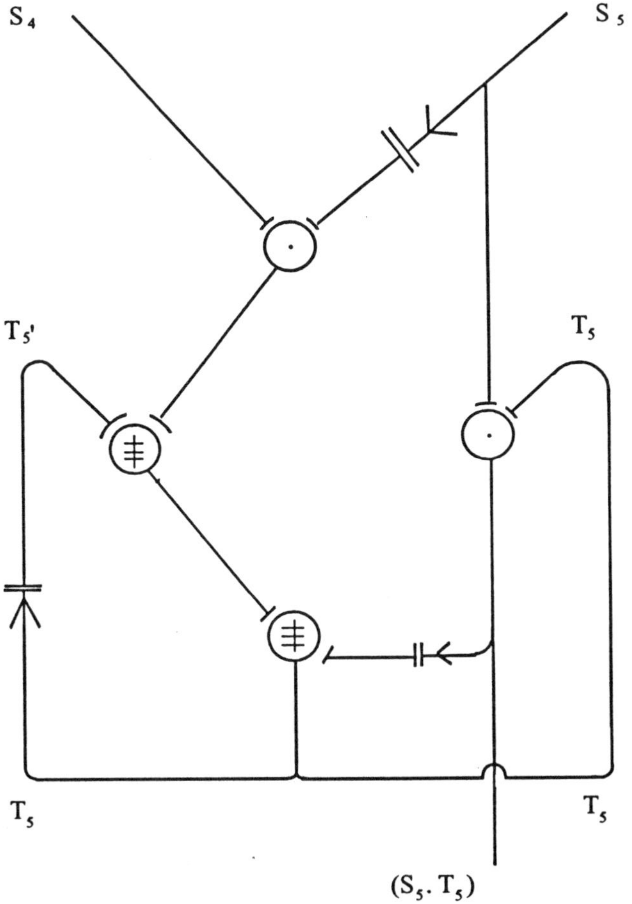

$$T_5 \equiv ((T_5' \ddagger (S_5' \,.\, S_4)) \ddagger (S_5 \,.\, T_5)')$$

Conclusion

It has been suggested that neurons are of three kinds, corresponding to the commutative and associative operations. The nervous connexions between them have been charted, using the formulae as a guide. We have in fact done no more than define certain variables in a different way. Given a formula it is possible to draw a matching diagram and *vice versa*.

The lines in the diagrams represent nerve fibres. Each is designated by a stimulus or trace, which is present when impulses are travelling along the fibre and absent when they are not.

A feature of the diagrams of the regular and intermittent traces is that they take the form of circuits. Each includes a device that delays transmission in order to produce an after-effect. Each also contains a neuron of variation which controls the flow of impulses within the circuit. The logic of this process, which we have called primary variation, is discussed in earlier pages.

Finally, a word about the timing of responses is opportune. We have said that logical operations as carried out in the brain may be regarded as instantaneous. But the concept of instantaneity is a relative one. It is only in contrast to the length of the space between a stimulus or trace and its after-effect that we may ignore the time taken to carry out an operation. That it must take some time is undeniable. However, the whole of the time between discharging receptors in the sense-organs and innervating effectors in the muscles is very much shorter than a space. What matters is the response to a cue must come before the after-cue — and well before it to affect the outcome.

Chapter VII

AN ALTERNATIVE NEUROLOGY

Logic of the Neuron

A more sophisticated way of drawing diagrams of the nervous system is given below. The unit is once again a single neuron.

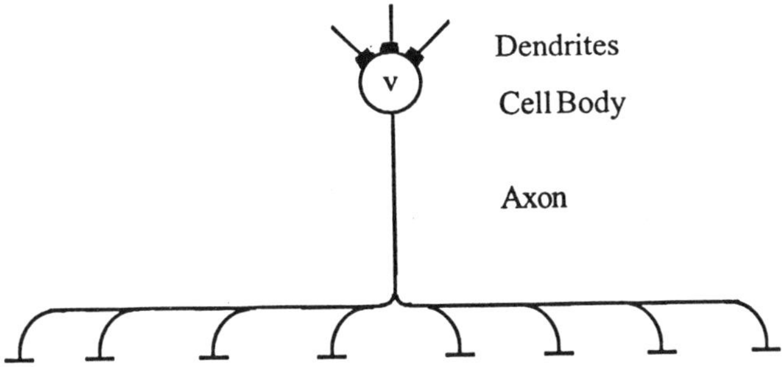

The only kind of neuron in this system is one of disjunction. Incoming fibres impinge on synaptic connexions on the dendrites and axon in order. The synapses are either excitatory or inhibitory.

$(S_1 \,.\, S_2)$ $(S_1 \not\supset S_2)$ $(S_2 \,.\, S_1)$ $(S_2 \not\supset S_1)$

The method is ideal for depicting sensory and motor interaction. Whereas in the previous chapter each excitation required one neuron and each inhibition two, in the alternative notation above all the excitations and inhibitions involve a single neuron. A much more economical deployment of neurons is thus possible. In this kind of arrangement motor interaction is a continuation of sensory interaction. The table below makes the point clear. The effective cues are simply defined in a different way.

Ta3	Ta2	Ta1	
0	0	0	(Sa . Ta[0]) ≡ (((Sa ꝥ Ta3) ꝥ Ta2) ꝥ Ta1)
0	0	1	(Sa . Ta[1]) ≡ (((Sa ꝥ Ta3) ꝥ Ta2) . Ta1)
0	1	0	(Sa . Ta[2]) ≡ (((Sa ꝥ Ta3) . Ta2) ꝥ Ta1)
0	1	1	(Sa . Ta[3]) ≡ (((Sa ꝥ Ta3) . Ta2) . Ta1)
1	0	0	(Sa . Ta[4]) ≡ (((Sa . Ta3) ꝥ Ta2) ꝥ Ta1)
1	0	1	(Sa . Ta[5]) ≡ (((Sa . Ta3) ꝥ Ta2) . Ta1)
1	1	0	(Sa . Ta[6]) ≡ (((Sa . Ta3) . Ta2) ꝥ Ta1)
1	1	1	(Sa . Ta[7]) ≡ (((Sa . Ta3) . Ta2) . Ta1)

The diagram which follows shows first the formation of these effective cues and second the stimuli of action.

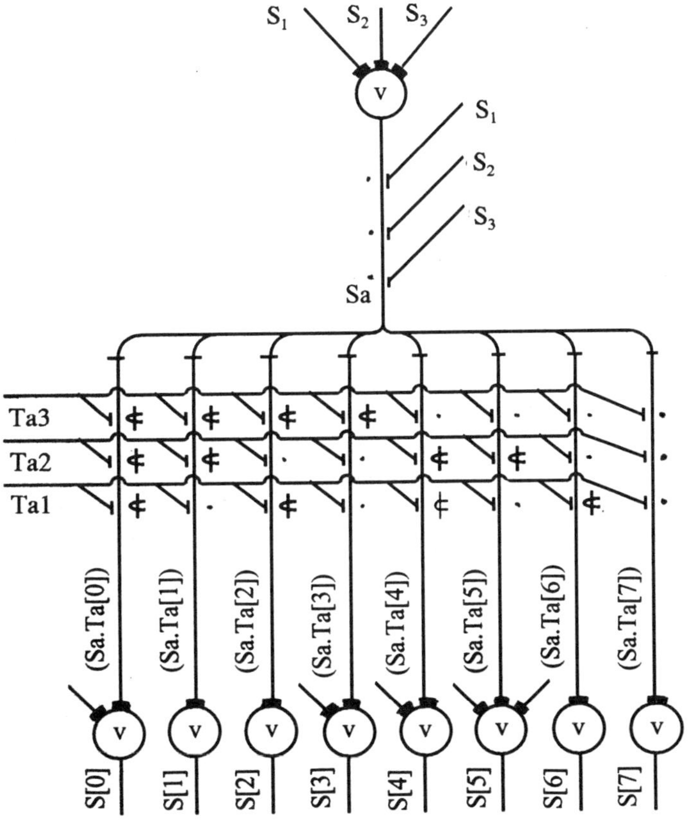
S1
S2
S3
v
S1
S2
S3
Sa
Ta3
Ta2
Ta1
(Sa.Ta[0])
(Sa.Ta[1])
(Sa.Ta[2])
(Sa.Ta[3])
(Sa.Ta[4])
(Sa.Ta[5])
(Sa.Ta[6])
(Sa.Ta[7])
v
v
v
v
v
v
v
v
S[0]
S[1]
S[2]
S[3]
S[4]
S[5]
S[6]
S[7]

Motor interaction takes place in the dendrites of the eight neurons above. The first synapse in the dendrite is depicted thus: ǂ

The system uses one neuron for each cue and one for each stimulus of action. In this book the number of possible actions (18) is greater than the number of cues (7) but this is not typical of human beings or animals. The number of possible actions open to the animal in comparison to cues is small and may be disregarded in estimating the approximate number of elements to which it is sensitive, so a nervous system of N elements requires rather more than 2^N neurons. The number of neurons in the human brain, for example, leads us to believe N is about 36 or 37. 2^N is then about a hundred thousand million.

Alternative formation of the Traces

The original definitions of regular and intermittent traces for the cue Sa in Chapter III were as follows.

Ta1 ≡ (Tá1 ǂ (Sá . Ŝ))

Ta2 ≡ (Tá2 ǂ (Tá1 þ Ta1))

Ta3 ≡ (Tá3 ǂ (Tá2 þ Ta2))

We have seen in Chapter I that variation can be defined in terms of disjunction and occultation. Theorems of regular and intermittent traces may be proved using that definition. The diagram which follows shows how these traces are formed by the alternative method. The theorems on which it is based are written beneath.

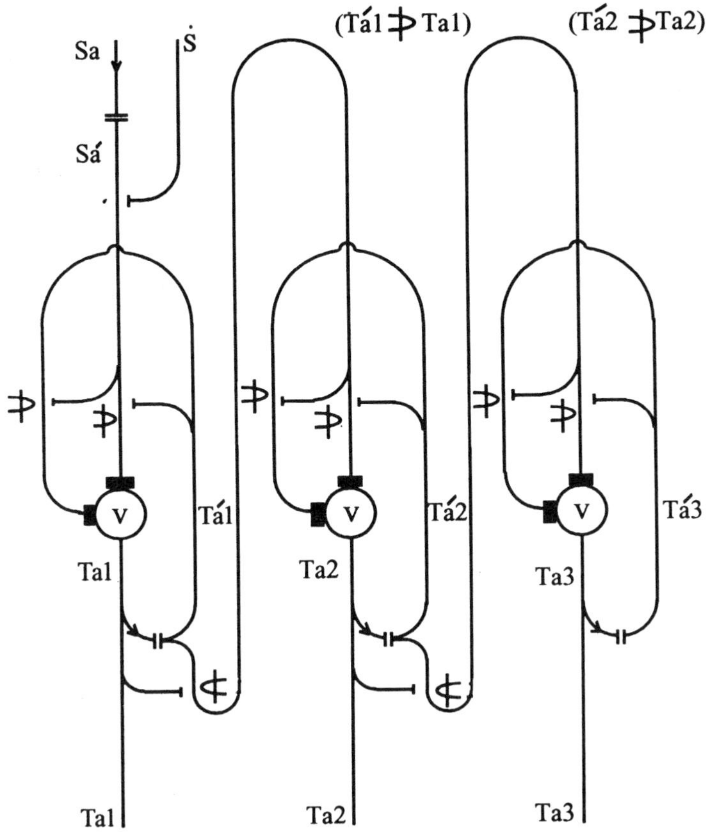

$$Ta1 \equiv ((Ta'1 \not\supset (Sa' \,.\, \dot{S})) \vee ((Sa' \,.\, \dot{S}) \not\supset Ta'1))$$

$$Ta2 \equiv ((Ta'2 \not\supset (Ta'1 \not\supset Ta1)) \vee ((Ta'1 \not\supset Ta1) \not\supset Ta'2))$$

$$Ta3 \equiv ((Ta'3 \not\supset (Ta'2 \not\supset Ta2)) \vee ((Ta'2 \not\supset Ta2) \not\supset Ta'3))$$

An Alternative Form of the Conditioning Trace

The definition of a conditioning trace is reproduced below.

$$T_5 \equiv ((T'_5 \not\equiv (S'_5 \,.\, S_4)) \not\equiv (S_5 \,.\, T_5)')$$

It can be expressed in terms of the alternative system as follows.

$$T_5 \equiv ((S_4 \,.\, S_5') \vee (T_5' \not\supset S_5'))$$

The diagram which follows shows the formation of this trace. It may be compared to the diagram of the same trace at the end of the previous chapter.

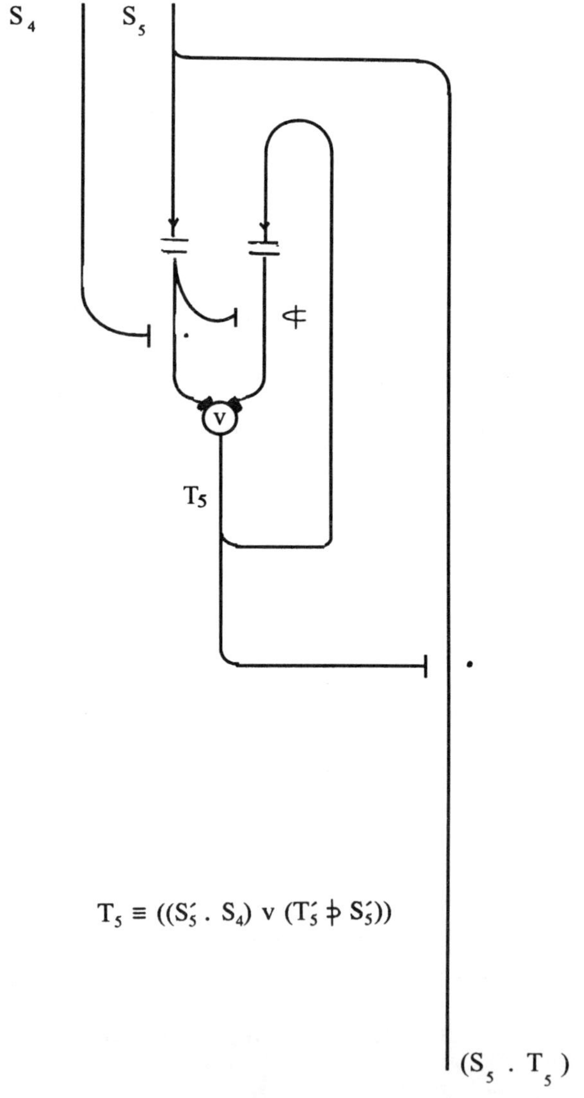

Conclusion

So far as intelligence is concerned diagrams of the formation of common regular and intermittent traces and their interaction could also be drawn in the alternative system. Motor interaction would again be a continuation of sensory interaction, with the comprehensive traces omitted, and this means the diagrams would be a complicated though accurate picture of the distribution of neurons and their connections in the animal brain.

This then completes our study of learning in animals. The theory proposed is an integrative one. We were able to show by symbolic logic that only one cue is present at once and that only one response ensues. Just as the cue is made of a pattern of elements so the response consists of a vast pattern of muscle-fibre contractions. The logic is reflected in the structure of the neurons, their terminal fibres reaching out in relays to a huge number of muscle-fibre effectors, bringing about a series of holistic actions.

The system illustrated in the previous chapter is easier to grasp, though its use of neurons in sensory and motor interaction is wasteful. The logical operation of variation may however take place at a primitive level in the brain, underpinning the formation of traces. And there are four more theoretical systems than the two we have pictured, each typified by definitions of traces that include variation. A logical psychology can only tell us the possibilities, in the shape of these different definitions. Psychology is a search for the rationale of behaviour and must find its ultimate corroboration in neurology, which alone can reveal the nature of the real nervous system.

The specification of the brain must be exact. There are an enormous number of tiny connecting neural strands and the length of each one is crucial. In this respect all the diagrams are of course rough approximations.

Appendix 1

THE ADAPTIVENESS OF INTELLIGENCE

Generalization ensures that only the most recent and therefore reliable experience is used in guiding an animal's behaviour. Compare the performance of animals where responses to the cues Sd and Se are linked (as in the theory) or not.

		1	2	3	4	5	6	7
LINKED	Sd or Se elicits	*	[10]	*	*	[14]	*	[11]
	$\dot{S}$(pain or need)	0	0	0	0	0	1	0

		1	2	3	4	5	6	7
NOT LINKED	Sd or Se elicits	*	[10]	*	*	[14]	*	[10]
	$\dot{S}$(pain or need)	0	0	0	0	0	1	0

The series above differ in space 7. From them we can see that in the case of linked responses the animal uses information obtained in space 6 while with responses which are not linked it has to rely on that gained earlier in space 3 and the aversive outcome of response [14] is ignored. If responses [10] and [14] really do converge the survival-value of linking them is obvious. Moreover, in the case of linked repertoires when information is pooled fewer nerve circuits are necessary. In this respect the most highly developed brains are the simplest.

Appendix 2

THE LOGIC OF BEHAVIOUR

The following meanings are frequently ascribed to the signs of contrariety, entailment and equivalence.

$|$ Not both ..... and .....

$\supset$ If .....then .....

$\equiv$ If and only if ..... then .....

Any of these relations may exist between pairs of stimuli which innervate effectors — at a microscopic level individual muscle-fibres. These stimuli are each equivalent to a disjunction (or variation) of stimuli of action. The reader will recall that the latter are contrary. The following theorems serve as examples.

$$(S[6] \vee S[15]) \mid (S[0] \vee S[7] \vee S[14])$$

$$(S[1] \vee S[6]) \supset (S[1] \vee S[4] \vee S[6] \vee S[10])$$

$$(S[2] \vee S[5]) \equiv (S[2] \vee S[5])$$

In the case of contrariety none of the stimuli of action on the left appear on the right. In the case of entailment all stimuli of action on the left also appear on the right. In the case of equivalence all stimuli of action on the left also appear on the right and *vice versa*.

Sometimes a stimulus which innervates an effector is equivalent to a single stimulus of action. Further examples of the three relations illustrate this point.

S[0]	|	(S[7] v S[10])
S[14]	|	S[2]
S[3]	⊃	(S[1] v S[3] v S[12])
S[11]	⊃	S[11]
S[9]	≡	S[9]

On the other hand, the stimuli (S[8] v S[12]) and (S[12] v S[13]) are neither contrary nor equivalent nor does one entail the other. They are said to be independent, in the sense that the presence or absence of one does not entail the presence or absence of the other. Stimuli which are contrary are not independent since the presence of one entails the absence of the other.

So either the relation of contrariety, entailment or equivalence holds good between pairs of stimuli which innervate effectors or no relation holds good between them, in which case they are independent. If no relation exists between them at least three stimuli of action must be involved, as for example (S[8] v S[12]) and (S[12] v S[13]).

The diagrams below illustrate the four possibilities.

Relation of Contrariety

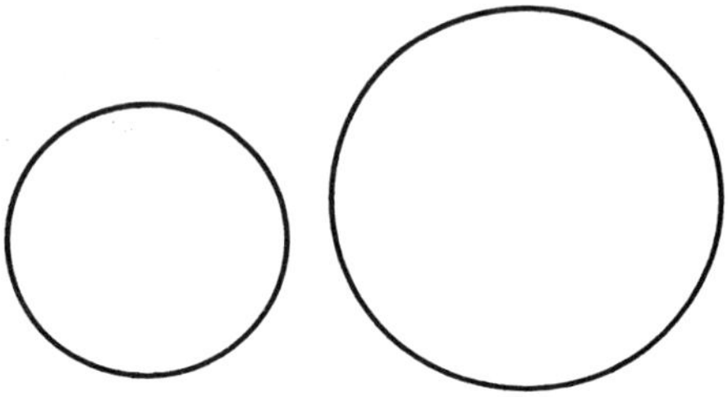

Relation of Entailment

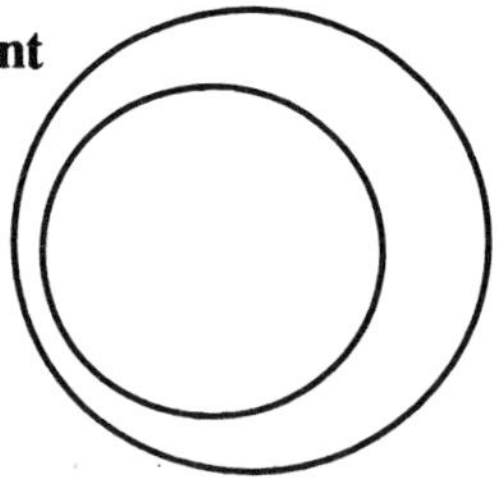

Relation of Equivalence

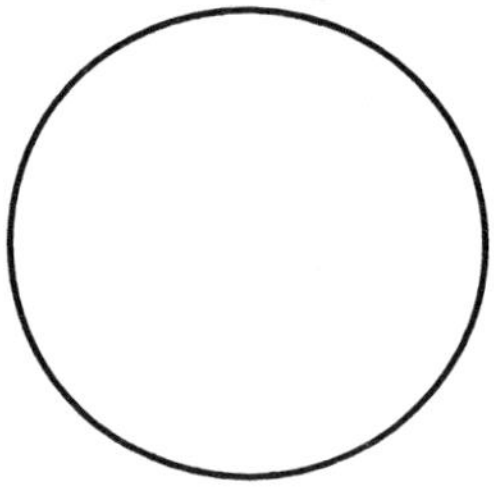

Independence

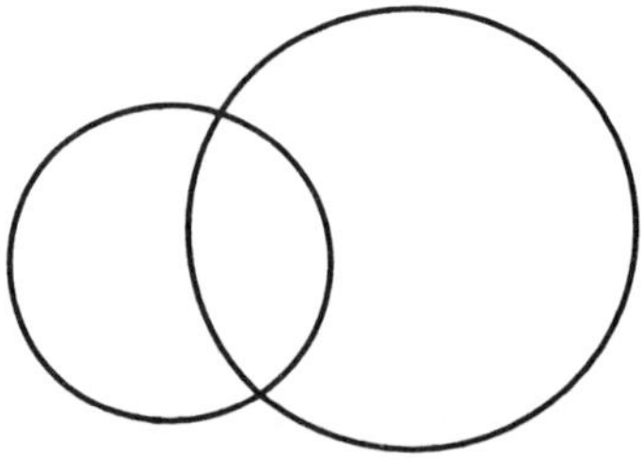

In the diagram of contrariety a point may lie in one circle or the other or in neither. It is impossible for it to lie in both circles. This is plainly an analogy of the relation of contrariety, where it is impossible for a pair of stimuli to be both present at once.

Again, in the diagram of entailment if a point lies in the smaller circle it must also lie in the larger. But it may lie in the larger circle and not the smaller. The third possibility is that it lies outside both circles. The analogy here covers the relation of entailment.

The next diagram, corresponding to equivalence, shows two circles with the same diameter and the same centre. The relation is an extreme case of entailment. A point must lie in both circles or neither.

The diagram of independence shows the intersection of two circles. A point may lie in both or in only one or in neither. This analogy applies to the case when no relation holds good between the two stimuli.

The four diagrams cover all possibilities. Just as pairs of stimuli which innervate effectors fall into four categories so it is impossible to draw more than four diagrams to match them.

The logic of relations between stimuli is the logic of behaviour. We may say of any two fibres in the musculature at any give time:

either (1) they cannot both contract
or (2) if one contracts the other does too
or (3) if and only if one contracts the other does too
or (4) their contractions are independent.

Appendix 3

NUMBERS

While our main intention has been to place psychology on a logical rather than a mathematical footing, the fact remains that mathematics is ultimately founded on logic and numerical concepts can often be defined in terms of logical ones. Thus, using the operations of conjunction and disjunction, we may write expressions for configurational stimuli which are only present when at least one, at least two and all three of the elements S_1, S_2 and S_3 are present respectively.

$(S_1 \vee S_2 \vee S_3)$	At least one
$((S_1 \,.\, S_2) \vee (S_1 \,.\, S_3) \vee (S_2 \,.\, S_3))$	At least two
$(S_1 \,.\, S_2 \,.\, S_3)$	All three

It is also possible to define exact whole numbers in the field of three elements by using the sign of occultation to join adjacent configurational stimuli above.

$((S_1 \vee S_2 \vee S_3) \not\supset ((S_1 \,.\, S_2) \vee (S_1 \,.\, S_3) \vee (S_2 \,.\, S_3)))$	One
$(((S_1 \,.\, S_2) \vee (S_1 \,.\, S_3) \vee (S_2 \,.\, S_3)) \not\supset (S_1 \,.\, S_2 \,.\, S_3))$	Two
$((S_1 .\, S_2 \,.\, S_3) \not\supset 0)$	Three

One of the elements is equivalent to at least one but not two or more, while two are equivalent to at least two but not three.

It is impossible to define zero — a number which is present when all elements in the field are absent — in our restricted system or in any other which excludes the contradictive function. Zero would be equivalent to the cue Sh which cannot be defined without the constant of presence 1. In a complete logic the definition would be written thus.

$$(1 \not\supset (S_1 \vee S_2 \vee S_3)) \qquad \text{Zero}$$

Although the constant of presence cannot be defined in our logic it may be shown by extrapolating values that a conjunctive association containing no members is invariably equivalent to 1. Similarly if a disjunctive association has no members it is always equivalent to 0. Meaning can then be ascribed to the two mechanisms pictured below in which there are no incoming fibres.

Impulses are produced spontaneously in the case of the neuron on the left. The mechanism could take the place of the receptacle and remove the anomaly of instinctive inaction. The neuron on the right does not appear to have any practical application.

In the expression for the number zero above the constant 1 represents the positive association of the cue Sh which has no members while in the expression for the number three the constant 0 represents the negative association of the cue Sa which has no members.

The two tables below show how values of presence and absence of conjunctions and disjunctions with no members are demonstrated by extrapolation.

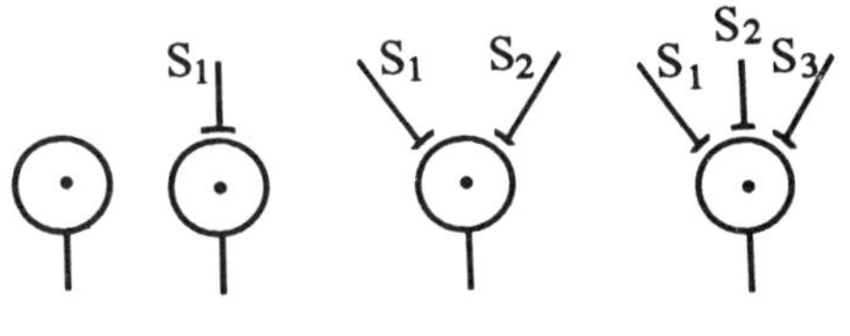

S_1	S_2	S_3				
1	1	1	1	1	1	1
0	1	1	1	0	0	0
1	0	1	1	1	0	0
0	0	1	1	0	0	0
1	1	0	1	1	1	0
0	1	0	1	0	0	0
1	0	0	1	1	0	0
0	0	0	1	0	0	0

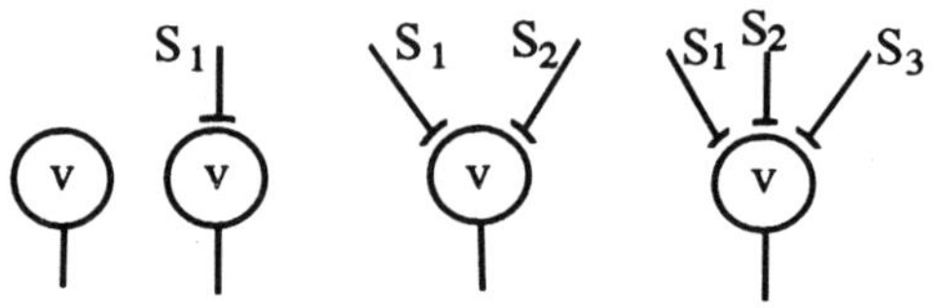

S_1	S_2	S_3				
1	1	1	0	1	1	1
0	1	1	0	0	1	1
1	0	1	0	1	1	1
0	0	1	0	0	0	1
1	1	0	0	1	1	1
0	1	0	0	0	1	1
1	0	0	0	1	1	1
0	0	0	0	0	0	0

Appendix 4

TUNING THEOREMS

Sometimes fibres divide and impinge two or more times on the same neuron. In the associational system depicted in Chapter VI tuning theorems describe this mechanism.

$$S_1 \equiv (S_1 \;.\; S_1)$$

$$S_1 \equiv (S_1 \vee S_1)$$

From these two theorems, which may be tested by drawing up tables of values, progressively more complex mechanisms can be described.

$$S_1 \equiv (S_1 \;.\; S_1)$$

$$S_1 \equiv (S_1 \;.\; S_1 \;.\; S_1)$$

$$S_1 \equiv (S_1 \;.\; S_1 .\; S_1 \;.\; S_1)$$

etc.

$$S_1 \equiv (S_1 \vee S_1)$$

$$S_1 \equiv (S_1 \vee S_1 \vee S_1)$$

$$S_1 \equiv (S_1 \vee S_1 \vee S_1 \vee S_1)$$

etc.

The tuning theorem of variation is as follows.

$$S_1 \equiv (S_1 \not\equiv S_1 \not\equiv S_1)$$

From this theorem we may prove many others.

$$S_1 \equiv (S_1 \not\equiv S_1 \not\equiv S_1)$$

$$S_1 \equiv (S_1 \not\equiv S_1 \not\equiv S_1 \not\equiv S_1 \not\equiv S_1)$$

$$S_1 \equiv (S_1 \not\equiv S_1 \not\equiv S_1 \not\equiv S_1 \not\equiv S_1 \not\equiv S_1 \not\equiv S_1)$$

etc.

In the mechanisms on the right the stimulus S_1 appears an odd number of times. Expressions containing an even number would be equivalent to the constant of absence.

The operational system described in Chapter VII is based on similar tuning theorems. A fibre may divide and impinge on more than one synapse on the same dendrite or axon.

$$(S_1 . S_2) \equiv ((S_1 . S_2) . S_2)$$

$$(S_1 \not\supset S_2) \equiv ((S_1 \not\supset S_2) \not\supset S_2)$$

More extended theorems can similarly be proved.

$$(S_1 . S_2) \equiv ((S_1 . S_2) . S_2)$$

$$(S_1 . S_2) \equiv (((S_1 . S_2) . S_2) . S_2)$$

$$(S_1 . S_2) \equiv ((((S_1 . S_2) . S_2) . S_2) . S_2)$$

etc.

$$(S_1 \not\supset S_2) \not\equiv ((S_1 \not\supset S_2) \not\supset S_2)$$

$$(S_1 \not\supset S_2) \equiv (((S_1 \not\supset S_2) \not\supset S_2) \not\supset S_2)$$

$$(S_1 \not\supset S_2) \equiv ((((S_1 \not\supset S_2) \not\supset S_2) \not\supset S_2) \not\supset S_2)$$

etc.

Appendix 5

THE LOGICAL PRIORITY OF INHIBITION

The operation underlying excitation, which is conjunction, may be defined in terms of occultation, which underlies inhibition.

$$(S_1 \;.\; S_2) \equiv (S_1 \not\supset (S_1 \not\supset S_2)) \quad \text{Def.}$$

The definition can be verified by the method of matrices. We have seen that the sign which is the converse of occultation may be defined in terms of it. This leads to two interesting formulae which may explain sensory and motor interaction. They are written below.

$$(S_1 \;.\; S_2) \equiv ((S_1 \not\subset S_2) \not\subset S_2)$$

$$(S_1 \not\supset S_2) \equiv ((S_1 \not\supset S_2) \not\supset S_2)$$

The expressions on the right bear a parallel relationship to each others. However, the operation of occultation alone is insufficient to define all five signs of the system. Variation or disjunction is also required.

THE SUPPRESSION OF PRIMARY FEELINGS

The Problem of Conflicting Needs

In my book *The Intelligence of Animals* I described a stimulus called the varying complex, symbolized by $\hat{S}$. It was defined as follows.

$$\hat{S} \equiv ((S_1 \nsupset S_2) \vee S_4 \vee (S_5 \,.\, T_5) \vee (S_6 \,.\, T_6) \vee (S_7 \nsupset (S_8 \vee (S_9 \,.\, T_9))))$$

Hunger	Pain	Fear(1)	Fear(2)	Thirst

Examples were given of two needs (which I shall here call hunger and thirst) pain, fear of pain and fear of fear of pain. Each of these stimuli elicits a primary feeling and entails the varying complex, in that way producing a change in the animal's future behaviour. The primary stimuli of reaction therefore make up a disjunction, as we see in the definition of $\hat{S}$ above. Clearly if the elements within it are completely independent primary variation will be a very frequent occurrence and learning will be largely impossible. To offset this difficulty I shall describe a mechanism which ensures that most of the elements above are not independent, so that primary variation only takes place when absolutely necessary.

Imagine a maze which contains both food and water but, as so often happens in nature, at different locations. Depending on whether they are hungry or thirsty, rats will have to make different sequences of responses to reach these goals. Suppose after training to do this, first under conditions of hunger and second of thirst, the animals are confronted with a situation in which they have been deprived of both food and water. According to the argument in my book the setting would now be a new one in which the rodents are required to start learning from scratch, having no previous experience under these circumstances. But in fact they will know what to do. Probably thirst will be satisfied first and hunger second. How do we describe the mechanisms behind this?

One solution that springs to mind is that the drive-stimuli of hunger and thirst are contrary elements, so that situations where both are present simply cannot arise. The two drive-stimuli in the formula above are S_1

and S_7. The drive-stimulus of hunger, which we shall take to be S_1, is contrary to the drive-stimulus of thirst, taken to be S_7, when they are composed as follows.

$$S_1 \equiv (S_{II} \not\supset S_I)$$

$$S_7 \equiv S_I$$

$$\therefore \quad S_1 \mid S_7$$

The inhibitory process above is called *suppression*. It ensures incidentally that the primary stimuli of reaction of the two needs are also contrary.

$$(S_1 \not\supset S_2) \mid (S_7 \not\supset (S_8 \vee (S_9 \,.\, T_9)))$$

In other words, so far as sensation is concerned, animals cannot feel both hunger and thirst at the same time, when this particular arrangement exists. When suppression covers a set of needs only one can be experienced at once. The needs are satisfied in order of importance.

The symbol S_i corresponds to a stimulus which is present when dehydration of cells in the body reaches a critical point, while S_{II} corresponds to a fall in the amount of glucose or amino-acids in the blood below an equally critical level. The actual mechanisms are certainly much more complex than this, but may be construed in logical terms along the same lines. For instance extreme hunger may take precedence over mild thirst. We are then dealing with three feelings of need: thirst and two intensities of hunger. Each is elicited by a different stimulus of reaction. Severe hunger depends on nutrients in the blood falling below a critical level even lower than that indicated by the presence of S_{II}.

Suppression of Pain

It has been found that rats will cross a lightly charged electric grid in order to satisfy a strong need. In the absence of the need the grid will be avoided. The mechanism of suppression is once again responsible for these results. In the varying complex above the element of pain (i.e. a mild shock) is represented by S_4. It is contrary to S_7 when made up as follows. The stimulus S_{III} is present when voltage reaches a certain level.

$$S_4 \equiv (S_{III} \not\supset S_I)$$

$$S_7 \equiv S_I$$

$$\therefore \quad S_4 \mid S_7$$

As in the case of the suppression of a need the stimuli of reaction are also contrary.

$$S_4 \mid (S_7 \not\supset (S_8 \vee (S_9 \, . \, T_9)))$$

Pain however has no suppressing effect on other feelings as a pain element entails the varying complex anyway.

Suppression of Fear

A neutral element that has acquired punishing power may also have a disruptive influence on the chaining of responses leading to a vital goal. As it is less strong than the pain it foreshadows there is even more reason why it should be suppressed. The mechanism behind the suppression of fear is obscure. It is a moot point whether the action of the neutral element or its conditioning trace is inhibited. Either way the primary feeling of fear (1) aroused by $(S_5 \, . \, T_5)$ is contrary to the drive-stimulus S_7 and feeling of thirst.

$$(S_5 \, . \, T_5) \mid (S_7 \not\supset (S_8 \vee (S_9 \, . \, T_9)))$$

The feeling of fear (2) aroused by $(S_6 \, . \, T_6)$ is also contrary to thirst, as we see in the theorem below.

$$(S_6 \, . \, T_6) \mid (S_7 \not\supset (S_8 \vee (S_9 \, . \, T_9)))$$

Possibly suppression ensures that the feelings of fear are contrary to hunger, which is a weaker need than thirst.

$$(S_5 \, . \, T_5) \mid (S_1 \not\supset S_2)$$

$$(S_6 \, . \, T_6) \mid (S_1 \not\supset S_2)$$

Summary

We can now draw up a list of the feelings in the varying complex above, in order of strength or importance.

Thirst

Pain

Hunger

Fear (1)

Fear (2)

Similar lists may be drawn up for other examples of the varying complex in which primary stimuli of reaction of different intensities of the same need appear, e.g. mild hunger, mild thirst, extreme hunger and extreme thirst. And it may be shown that other relations hold good between elements in these examples. It is likely that different intensities of the same need are satisfied by the same object, whether it be food or water. In other words the secondary stimuli of satisfaction are equivalent. Alternatively the object of one intensity of need may entail or be entailed by the object of another. The same is perhaps true of the drive-stimuli themselves. The drive-stimulus of a severe need may entail the drive-stimulus of a mild form of the same need. All in all there are many cases of elements in the complex being related, thus implying that a vast number of situations or cues are logically impossible.

RESPONSES TO PART OF THE OVERALL SITUATION

The main premise in my book *The Intelligence of Animals* was that an animal responds to the whole situation confronting it at any given moment. But the cue, as I called it, may in many cases be only a part of this situation. Consider two of the examples in the preface.

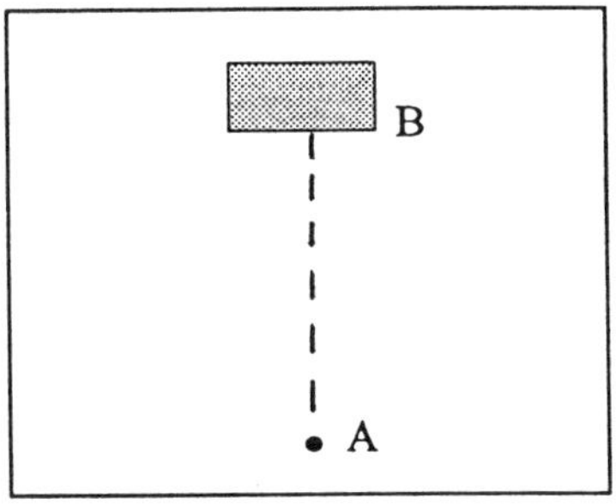

Fig.1

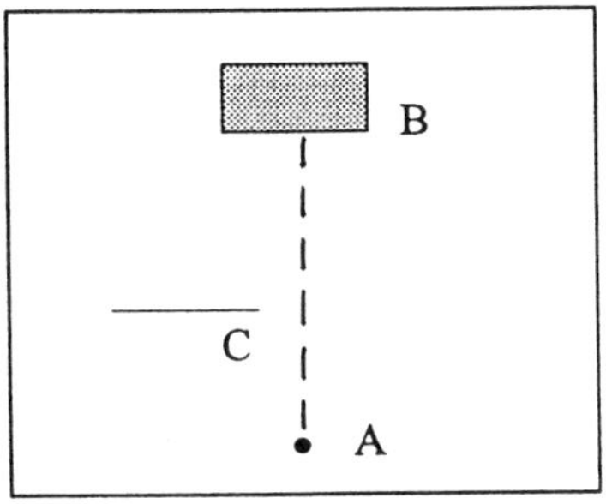

Fig.3

In Fig.1 the chicken learns to run to the tray of food B from A. In Fig.3 it again runs to the tray though the situation now includes the screen C. This phenomenon I called simple generalization. I assumed that the overall situations in Fig.1 and Fig.3 were different, as indeed they were, and that the response was made to their logical disjunction. But a better explanation would surely be that the chicken simply ignores the screen C in Fig.3, and in this respect responds to only a part of the situation.

If we recall that the argument in the book centred around a field of three elements S_1, S_2 and S_3, then the situation in Fig.1 might correspond to the cue Sg and the situation in Fig.3 to the cue Sc. Simple generalization would then mean the bird responds to the disjunction of the cues (Sc v Sg). The cues Sc and Sg were defined as follows.

$$Sc \equiv ((((S_1 \vee S_2 \vee S_3) \,.\, S_1) \not\supset S_2) \,.\, S_3)$$

$$Sg \equiv ((((S_1 \vee S_2 \vee S_3) \,.\, S_1) \not\supset S_2) \not\supset S_3)$$

The screen may be identified with the element S_3. Logical rules dictate that the disjunction of the cues Sc and Sg can be expressed most easily by the formula below, in which that element only figures in the receptacle.

$$(Sc \ v \ Sg) \equiv (((S_1 \ v \ S_2 \ v \ S_3) \ . \ S_1) \not\supset S_2)$$

The definition above shows that a 'cue' is not necessarily synonymous with a 'situation'. In this case the cue only consists of part of the situation facing the chicken. The bird as it were 'pays attention' to only the presence of S_1 and absence of S_2. The element S_3 is ignored.

Assuming there are no more instances of the same type in this new system, we may draw up a table of definitions of six cues as follows. A cue when reduced like (Sc v Sg) may be represented as Scg.

$$Sa \equiv ((((S_1 \ v \ S_2 \ v \ S_3) \ . \ S_1) \ . \ S_2) \ . \ S_3)$$

$$Sb \equiv ((((S_1 \ v \ S_2 \ v \ S_3) \not\supset S_1) \ . \ S_2) \ . \ S_3)$$

$$Scg \equiv (((S_1 \ v \ S_2 \ v \ S_3) \ . \ S_1) \not\supset S_2)$$

$$Sd \equiv ((((S_1 \ v \ S_2 \ v \ S_3) \not\supset S_1) \not\supset S_2) \ . \ S_3)$$

$$Se \equiv ((((S_1 \ v \ S_2 \ v \ S_3) \ . \ S_1) \ . \ S_2) \not\supset S_3)$$

$$Sf \equiv ((((S_1 \ v \ S_2 \ v \ S_3) \not\supset S_1) \ . \ S_2) \not\supset S_3)$$

$$0 \equiv ((((S_1 \ v \ S_2 \ v \ S_3) \not\supset S_1) \not\supset S_2) \not\supset S_3)$$

Each of the six cues above elicits a repertoire of responses, as they did in the system examined in my book. The example of simple generalization (Sc v Sf) given there cannot be reduced to one cue because Sc is the obverse of Sf. But when reduction can take place it is possible to draw diagrams of the nervous system in which the number of neurons is significantly smaller. It is another way in which the number of elements may be increased in ratio to the number of neurons in the brain. The human brain, for instance, is sensitive to considerably more than thirty-six or so truly independent elements, even though there are no more than between two to the power of thirty-six and two to the power of thirty-seven neurons within it.

SENSORY INTERACTION. ANOTHER ARRANGEMENT

In Chapter VII I described sensory interaction as a process of successive excitations and inhibitions on the axon of a sensory neuron. Another possibility follows.

(1) A single neuron of disjunction subserves the formation of the receptacle for all the cues.

(2) Sensory interaction occurs on one of the dendrites of a second neuron rather than on its axon.

(3) Simple generalization depends on interaction on more than one dendrite of this neuron. Reduction of cues also takes place at the same level, where possible, by using fewer dendrites.

Such a plan economizes with neurons in an even better way than that described in Chapter VII and my paper on reduction. In simple generalization fewer neurons are employed whether reduction is possible or not.

One of the implications of both schemes is that with reduction the number of independent elements and hence feelings can theoretically run into thousands or even millions.

ANOTHER EXAMPLE OF THE VARYING COMPLEX

It is possible to show the interplay of one need and two intensities of another, such as thirst and mild and extreme hunger, by formulating an alternative varying complex.

$$\hat{S} \equiv ((\underset{\text{Mild hunger}}{S_1} \not\supset \overset{\text{Food}}{S_2}) \vee (\underset{\text{Thirst}}{S_3} \not\supset \overset{\text{Water}}{S_4}) \vee (\underset{\text{Extreme hunger}}{S_5} \not\supset \overset{\text{Food}}{S_6}))$$

Let us assume the objects of the two forms of hunger are equivalent.

$$S_2 \equiv S_6$$

As we saw in the paper on suppression the next two theorems are true.

$$S_1 \mid S_3$$

$$S_3 \mid S_5$$

Finally the drive-stimulus of mild hunger is entailed by that of extreme hunger precisely because their two objects are the same. Food satisfies both degrees of deprivation.

$$S_5 \supset S_1$$

It is not essential for the drive-stimulus S_5 to be represented in the antecedent cue, apart from its inclusion in the receptacle. Only the drive-stimulus S_1 it entails must be found there. The element S_5 is eliminated by reduction. This is a general rule applying to all strong forms of the same need. Their drive-stimuli appear in the varying complex but not among the operators of the cue. Perhaps the term 'drive-stimulus' should be reserved for primary elements which do play a part before a response as well as after; that is to say S_1 rather than S_5. The idea of a 'drive' seems to suggest this, as something which precedes and impels activity. But this is only a semantic question.

All the theorems above depend for their validity on the logical composition of the elements they contain.

A MATHEMATICAL THEORY OF PREDICTION

Consider a series of values of presence or absence of a stimulus. In any place in that series let the probability be p that a value is presence. Then the probability is (1 - p) that the value is absence (i.e. not presence).

Let us compare two methods of prediction. In Method A if the value is presence at a given place in the series then we predict the next value will also be presence, while if a value is absence we predict the next value will also be absence. The chances that a correct prediction will be made is as follows.

$$p^2 + (1 - p)^2$$

In Method B if a value is presence at a given place in the series than we predict the next value will be absence, while if a value is absence we predict the next value will be presence. The chances that a correct prediction will be made is then as follows.

$$p(1 - p) + (1 - p)p$$

If the total probability that a correct prediction will be made under Method B is subtracted from the total probability under Method A the result is as follows.

$$(1 - 2p)^2$$

This total is never less than zero whatever the value of p. Hence A is never a worse method of prediction than B. It may also be shown that Method A is optimal. Moreover the probability of correct prediction under A is never less than ½ while that under B is never more than ½ (assuming p is constant). This proof supports the all-or-nothing laws of my theory.

INCONSISTENT BEHAVIOUR?

The all-or-nothing version of the law of effect, as exemplified by the 'carrot and stick' theory of learning, is an oversimplification. As far as the 'nothing' or role of the 'stick' is concerned, instinctive responses proved to be an exception. An instinctive response is one that is always repeated whether it is followed shortly by the presence of the varying complex or not.

There is another possible exception to the rule. A repertoire assigned to a cue may contain the same response more than once, and in some cases two or more times in succession. For example, a repertoire may consist of the four responses [1], [1], [2] and [3], in that order. Although it is difficult to see the survival-value of such an arrangement we cannot rule it out completely. It is probably a rare anomaly.

The 'all' part of the all-or-nothing law, or 'carrot', always holds good however. A response which is followed shortly by the absence of the varying complex will be repeated on the next trial in every case. When conditions are exactly the same the same response will ensue. This accords with a comparatively simple view of causation.

Reinforcement theorists introduced a principle of uncertainty into psychology to account for the unpredictability of animal and human behaviour. They maintained that a response followed by a reward would not necessarily be repeated, though it would become more likely. Similarly a response followed by punishment would become less likely, without its non-occurrence becoming certain. In this way they were able to explain how a punished response is sometimes repeated and, conversely, how a rewarded one is sometimes not.

In my theory a situation is defined much more stringently. So-called errors are explained by assuming cues in successive trials are slightly different. In the case of a response followed by an aversive outcome there is also the possibility the next one in the repertoire will not be exactly the same either, but only appear to be so. There are reasons too why in some cases a punished response really is repeated, as we saw above. So my theory can account for what seem to be the vagaries of behaviour quite as well as reinforcement theories. The all-or-nothing principle, though it does not invariably hold good, is in general a great improvement on the ones put forward by them. It is easier to describe the mechanisms behind it and has much greater survival-value.

The idea of a fixed repertoire, simplistic though it may sound at first, can be defended quite easily. Each response is tailor-made to fit the situation in hand. If the initial response in the repertoire is unsuccessful one which is next most likely to solve the problem will follow, taking into account the failure of the first, and so on. This would appear to be superior to a system based on random selection.

Repertoires are of course finite but there is no reason why they should be large anyway. A repertoire may be small because only a few responses fit that particular situation. And the cyclic nature of behaviour is probably often obscured because a very large number of factors are involved. We may also be unsure what an animal will do next as many of them are internal and out of sight. The huge number of situations which may confront the animal prevents us from knowing precisely how it will act in the future. This is the real reason why behaviour is so variable.

SUMMARY

Theories of learning, particularly those making use of the concept of reinforcement, tend to deal with probabilities. It is argued that a successful response to a situation is more likely to be repeated when the situation arises again. We proposed, on the other hand, that repetition under these conditions is certain. We were then able to describe the mechanisms which bring it about.

Such a position, attractive as it is on grounds of mechanical simplicity, can only be defended if the situation in question is defined stringently. Only if exactly the same one is presented again will the principle remain true. But the chances this will happen become increasingly remote as the number of distinguishable situations grows. Learning of any kind in the higher animals becomes well-nigh impossible. To meet this difficulty they rely on generalization. Groups of situations are associated in such a way that if a response to one of them is successful the effect is as if linked responses to the others had been successful too. When these situations arise such responses follow.

Generalization on this view is not simply a matter of making the same response to different situations, although this can happen. More often it means the production of entirely new and unrehearsed behaviour. But, paradoxically, this is still determined by experience. And it can also be explained in mechanistic terms, as we have seen.

So learning is more complicated than hitherto supposed. Our contention is that inborn beliefs about the possible consequences of untried actions play a vital role. The idea of convergence was introduced to show how they may be justified.

We have been able to define intelligence, contrasting it with simple trial and error. The greater the difference between converging responses (and, equally, the situations which evoke them) the more we think of this quality being involved.

LIST OF SYMBOLS

1	Value of stimulus presence
0	Value of stimulus absence. Constant of absence
S_1, S_2, S_3, etc.	Elements
Sa, Sb, Sc, etc.	Cues
S[0], S[1], S[2], etc.	Stimuli of action
$\dot{S}$	Varying complex
Ta1, Tb1.	Regular traces
Ta2, Tb2. Ta3.	Intermittent traces
Ta[0], Tb[5], etc.	Comprehensive traces
Tcf1, Tde1.	Common regular traces
Tde2	Common intermittent trace
Tcf[9], etc. Td[10]e[14], etc.	Common comprehensive traces
T_5, T_6, T_9.	Conditioning traces
S′	After-effect of a stimulus
T′	After-effect of a trace
≡	Relation of equivalence
|	Relation of contrariety
⊃	Relation of entailment
ǂ	Operation of variation
.	Operation of conjunction
v	Operation of disjunction
ꝥ	Operation of occultation